"The right to water remains elusive for a great number of people around the world. Despite decades of efforts by activists, policy-makers, and committed scholars, access to water remains deeply contested and unevenly distributed. This superb collection teases out why this is the case and, more importantly, presents a range of actions and principles, mobilised by a great variety of communities, that open possible pathways for a more just, democratic and egalitarian distribution of a key resource for securing livelihood. This is a must read for all those who still believe that a more humane, sustainable, and egalitarian access to the earth's waters is not only desirable, but necessary."
– *Professor Erik Swyngedouw, The University of Manchester, UK and Honorary Doctor of Roskilde University, Denmark and University of Malmö, Sweden*

"The world faces a growing water crisis. This is not just about water availability, but about distribution: who gets what and how water is used. Sultana and Loftus' book is ground-breaking. It provides a narrative of and pathways to water justice. It is a must read for anyone who cares water and our common future."
– *Professor R. Quentin Grafton, The Australian National University and the UNESCO Chair in Water Economics and Transboundary Water Governance.*

"This collection of essays provides much-needed intellectual inspiration for re-imagining water. Its clear message is that realizing the right to water involves re-organizing and re-thinking ways of relating to water, but also requires engaging with the wider transformations needed to make this world more sustainable and just."
– *Professor Margreet Zwarteveen, Professor of Water Governance, IHE Delft Institute for Water Education and University of Amsterdam, The Netherlands*

Water Politics

Scholarship on the right to water has proliferated in interesting and unexpected ways in recent years. This book broadens existing discussions on the right to water in order to shed critical light on the pathways, pitfalls, prospects, and constraints that exist in achieving global goals, as well as advancing debates around water governance and water justice.

The book shows how both discourses and struggles around the right to water have opened new perspectives, politics and possibilities in water governance, fostering new collective and moral claims for water justice, while effecting changes in laws and policies around the world. In light of the 2010 UN ratification on the human right to water and sanitation, shifts have taken place in policy, legal frameworks, local implementation, as well as in national dialogues. Chapters in the book illustrate the novel ways in which the right to water has been taken up in locations drawn globally, highlighting the material politics that are enabled and negotiated through this framework in order to address ongoing water insecurities. This book reflects the urgent need to take stock of debates in light of new concerns around post-neoliberal political developments, the challenges of the Anthropocene and climate change, the transition from the Millennium Development Goals (MDGs) to the Sustainable Development Goals (SDGs), as well as the mobilizations around the right to water in the global North.

This book is essential reading for scholars and students of water governance, environmental policy, politics, geography, and law. It will be of great interest to policymakers and practitioners working in water governance and the human right to water and sanitation.

Farhana Sultana is Associate Professor in the Department of Geography at Syracuse University, Syracuse, New York, USA.

Alex Loftus is Reader in the Department of Geography at King's College London, UK.

Earthscan Water Text

The International Law of Transboundary Groundwater Resources
Gabriel Eckstein

Water Governance and Collective Action
Multi-scale challenges
Edited by Diana Suhardiman, Alan Nicol and Everisto Mapedza

Water Stewardship and Business Value
Creating Abundance from Scarcity
William Sarni and David Grant

Equality in Water and Sanitation Services
Edited by Oliver Cumming and Thomas Slaymaker

Environmental Health Engineering in the Tropics
Water, Sanitation and Disease Control (third edition)
Sandy Cairncross and Sir Richard Feachem

Flood Risk Management
Global Case Studies of Governance, Policy and Communities
Edited by Edmund Penning-Rowsell and Matilda Becker

Water Ethics
A Values Approach to Solving the Water Crisis (second edition)
David Groenfeldt

The Water Footprint of Modern Consumer Society
(second edition)
Arjen Y. Hoekstra

Water Politics: Governance, Justice and the Right to Water
Edited by Farhana Sultana and Alex Loftus

For more information about this series, please visit: https://www.routledge.com/Earthscan-Water-Text/book-series/ECEWT

Water Politics
Governance, Justice and the Right to Water

Edited by
Farhana Sultana and Alex Loftus

First published 2020
by Routledge
2 Park Square, Milton Park, Abingdon, Oxon OX14 4RN

and by Routledge
52 Vanderbilt Avenue, New York, NY 10017

Routledge is an imprint of the Taylor & Francis Group, an informa business

© 2020 selection and editorial matter, Farhana Sultana and Alex Loftus; individual chapters, the contributors

The right of Farhana Sultana and Alex Loftus to be identified as the authors of the editorial material, and of the authors for their individual chapters, has been asserted in accordance with sections 77 and 78 of the Copyright, Designs, and Patents Act 1988.

All rights reserved. No part of this book may be reprinted or reproduced or utilized in any form or by any electronic, mechanical, or other means, now known or hereafter invented, including photocopying and recording, or in any information storage or retrieval system, without permission in writing from the publishers.

Trademark notice: Product or corporate names may be trademarks or registered trademarks and are used only for identification and explanation without intent to infringe.

British Library Cataloguing in Publication Data
A catalogue record for this book is available from the British Library

Library of Congress Cataloging-in-Publication Data
A catalog record has been requested for this book

ISBN: 978-1-138-32002-4 (hbk)
ISBN: 978-1-138-32003-1 (pbk)
ISBN: 978-0-429-45357-1 (ebk)

Typeset in Sabon
by Taylor & Francis Books

Contents

List of tables ix
List of contributors x
Foreword xiii
Preface xvi

1 The right to water in a global context: challenges and transformations in water politics 1
   FARHANA SULTANA AND ALEX LOFTUS

2 Valuing water: rights, resilience, and the UN High-Level Panel on Water 15
   JEREMY J. SCHMIDT

3 Making space for practical authority: policy formalization and the right to water in Mexico 28
   KATIE MEEHAN

4 Turning to traditions: three cultural-religious articulations of fresh waters' value(s) in contemporary governance frameworks 42
   CHRISTIANA ZENNER

5 The right to bring waters into being 54
   JAMIE LINTON

6 The rights to water and food: exploring the synergies 68
   LYLA MEHTA AND DANIEL LANGMEIER

7 Water-security capabilities and the human right to water 84
   WENDY JEPSON, AMBER WUTICH AND LEILA M. HARRIS

8 Rights on the edge of the city: realizing of the right to water in informal settlements in Bolivia 99
   ANNA WALNYCKI

9 Human right to water and bottled water consumption: governing at the intersection of water justice, rights and ethics 113
RAÚL PACHECO-VEGA

10 Against the trend: structure and agency in the struggle for public water in Europe 129
ANDREAS BIELER

11 Remunicipalization and the human right to water: a signifier half full? 143
DAVID A. MCDONALD

12 Citizen mobilization for water: the case of Thessaloniki, Greece 161
JERRY VAN DEN BERGE, RUTGERD BOELENS AND JEROEN VOS

13 Race, austerity and water justice in the United States: fighting for the human right to water in Detroit and Flint, Michigan 175
CRISTY CLARK

14 Class, race, space and the "right to sanitation": the limits of neoliberal toilet technologies in Durban, South Africa 189
PATRICK BOND

Index 203

Tables

3.1 Policy instruments applicable to the human right to water in Mexico 33
9.1 Two vignettes that exemplify private actors substituting public action towards enabling the human right to water 121

Contributors

About the Editors

Alex Loftus is Reader in Political Ecology at King's College London, UK, where he also leads the Contested Development research domain. His research considers the political ecology of water with a particular focus on the politics of infrastructure, financialization, and struggles for water justice.

Farhana Sultana is Associate Professor of Geography at the Maxwell School of Citizenship and Public Affairs at Syracuse University, USA, where she is also the Research Director (Environment) in the Program for the Advancement of Research on Conflicts and Collaboration (PARCC). Her research interests are broadly in water governance, climate justice, urban citizenship, and postcolonial development.

Contributors

Andreas Bieler is Professor of Political Economy and Fellow of the Center for the Study of Social and Global Justice (CSSGJ) in the School of Politics and International Relations at Nottingham University, UK. Together with Adam David Morton, he is the author of "Global Capitalism, Global War, Global Crisis" (Cambridge University Press, 2018).

Rutgerd Boelens is Professor "Water Governance and Social Justice", Wageningen University, Netherlands; Professor "Political Ecology of Water", Centre for Latin American Research and Documentation (CEDLA)/University of Amsterdam; Visiting Professor Catholic University Peru and Central University Ecuador. He directs the Justicia Hídrica/Water Justice alliance (www.justiciahidrica.org). Recent books: "Water Justice" (Cambridge University Press, 2018); "Water, Power, and Identity" (Routledge, 2015).

Patrick Bond is Professor of Political Economy in the Wits School of Governance at the University of the Witwatersrand, South Africa, and honorary professor at the University of KwaZulu-Natal where from 2004 until 2016 he directed the Center for Civil Society. His research and applied work combine political economy and political ecology.

List of contributors xi

Cristy Clark is a Lecturer at the Southern Cross University School of Law and Justice, Australia, where she teaches human rights law. Her research focuses on legal geography, the commons, and the intersection of human rights, neoliberalism, activism, and the environment.

Leila M. Harris is Professor at the University of British Columbia, Canada. Her work examines social, cultural, political-economic, institutional, and equity dimensions of environmental and resource issues, with research focused on water politics, water governance shifts, water insecurity, the human right to water, and considerations important for environmental justice.

Léo Heller is the UN Special Rapporteur on the human rights to safe drinking water and sanitation (since 2014) and also a researcher in the Oswaldo Cruz Foundation, Brazil. Previously, he was Professor in the Department of Sanitary and Environmental Engineering at the Federal University of Minas Gerais, Brazil.

Wendy Jepson is University Professor of Geography at Texas A&M University, USA. She leads several water security research projects, funded by the National Science Foundation and the Texas A&M Presidential Excellence Fund, that address water governance, water security, and environmental justice for the benefit of advancing community and human well-being.

Daniel Langmeier has a BSc in Agricultural Science from the ETH Zurich and an MA in Development Studies from the Institute of Development Studies (IDS), UK. He is the coordinator of the human rights organization Honduras Forum Switzerland, with which he works together with campesino groups and indigenous peoples at the intersection of the right to food and the right to water.

Jamie Linton leads a research program on rivers at the Université de Limoges in France. He is the author of "What Is Water? The History of a Modern Abstraction" (University of British Columbia Press, 2010).

David A. McDonald is Professor of Global Development Studies at Queen's University, Canada, and Director of the Municipal Services Project. He is the editor of several books, including "Making Public in a Privatized World: The Struggle for Essential Services" (Zed Books, 2016) and "Alternatives to Privatization: Public Options for Essential Services in the Global South" (Routledge, 2012).

Katie Meehan is a Senior Lecturer in Human Geography at King's College London, UK. Her interests are in cities, infrastructure, water governance, informal urban development, environmental justice, and the politics of climate change knowledge.

Lyla Mehta is a Professorial Fellow at the Institute of Development Studies, UK. She uses the case of water and sanitation to focus on the politics of human rights, gender, scarcity, access to resources, resource grabbing, power, and policy processes. She was Team leader of the UN High Level Panel of Experts on Food Security and Nutrition (HLPE) Report on Water for Food Security and Nutrition (2014–2015).

xii *List of contributors*

Raúl Pacheco-Vega is an Assistant Professor in the Public Administration Division of the Center for Economic Teaching and Research (Centro de Investigacion y Docencia Economicas, CIDE) in Mexico. He studies environmental politics, sanitation and water governance, solid waste management, neo-institutional theory, transnational environmental social movements, and research methods.

Jeremy J. Schmidt is Assistant Professor of Geography at Durham University, UK. He is the author of "Water: Abundance, Scarcity, and Security in the Age of Humanity" (New York University Press, 2017) and the co-editor of "Water Ethics: Foundational Readings for Students and Professionals" (Island Press, 2010).

Jerry van den Berge is consultant in water management in the Netherlands. He was policy advisor on water, energy, and sustainable development in the European Federation of Public Service Unions (EPSU, Brussels) and the coordinator of the first successful European Citizens' Initiative "Water is a human right!", popularly known as "Right2Water".

Jeroen Vos is Assistant Professor in the department of Water Resources Management of Wageningen University, the Netherlands. As a water policy advisor, he worked for a decade in Peru and Bolivia. His current research interests are the dynamics and discourses of water use by agribusinesses in Latin America.

Anna Walnycki is a senior researcher in the Human Settlements Group at the International Institute for Environment and Development (IIED), UK. She works with grassroots organizations in cities across the global South to understand how basic services intersect with informal processes of urbanization.

Amber Wutich is President's Professor of Anthropology and Director of the Center for Global Health at Arizona State University, USA. She directs the Global Ethnohydrology Study, a large cross-cultural study of water knowledge and management. Her two decades of community-based fieldwork explore how inequitable and unjust water institutions shape human wellbeing.

Christiana Zenner is Associate Professor in the Department of Theology at Fordham University, USA. She is the author of "Just Water: Theology, Ethics, and Global Water Crises" (Orbis Books, 2018) and works on questions pertaining to ethics and the values of fresh waters.

Foreword

I am pleased to welcome the book organized by Sultana and Loftus. Their previous work – *The right to water: politics, governance and social struggles* – published in 2012, just after the recognition of the Human Rights to Water and Sanitation (HRtWS) by the UN General Assembly and the Human Rights Council, has played an important role in instigating intellectual and political discussions on the meaning of those rights. Correctly, they addressed tensions and possibilities, which helped to confirm the idea that, although key, the legal provision of the HRtWS alone is not sufficient to guarantee their realization nor to protect people living in vulnerable situations. They also stressed the risk that groups of interest in the water sector frame processes of water commodification and social exclusion as rights realization.

This book, published at a very timely moment – nine years after the first UN resolution – updates the debate from the perspective of the social sciences, particularly that of Critical Geography, through case studies and theoretical reflections. The exercise of articulating debates on the HRtWS together with other frameworks, such as water justice and water governance, as well as with the assessment of struggles of civil society for the realization of the rights in several countries, is a great added-value of the book. Comparing this book with the 2012 edition – perhaps through the biased perspective of someone who needs to be optimistic about the potentialities of the human rights framework – the current work acknowledges more clearly that the HRtWS are a powerful tool for a better future for the scenario of access to water services, providing a legal and political basis for diverse actors to struggle for them. However, I leave it to each reader to draw their own conclusion.

My experience as the UN Special Rapporteur on the human rights to safe drinking water and sanitation since December 2014, particularly in undertaking country visits, has revealed the potentialities of this framework as an analytical lens for assessing policies, institutional arrangements, and the current, real circumstances of access to services. This lens is not a static one but has also a potential for change, as pressure on governments for the progressive realization of those rights. At the same time, this experience has revealed that the major challenges are still in place and impede the true incorporation of these rights in the political arena due to a variety of reasons. Sometimes, it is more comfortable for political actors to manage services through business as usual, in a technocratic way. Sometimes, there is really a lack of awareness about the HRtWS framework, from both governmental actors and/or from civil society organizations addressing water issues at national levels. The latter, more often than not, are simply absent or often focused on service implementation. Obviously, there are sound exceptions to this, such as those situations explored in the chapters of this volume addressing cases across countries.

One of the most stimulating activities of the mandate of the UN Special Rapporteur is the preparation of thematic reports, which aim to provide an interpretation of aspects of the HRtWS and to develop further issues still poorly understood in this regard. This is an intellectual endeavor, from the selection of the theme to the careful preparation of the report. In identifying the gaps of knowledge for the report themes (a total of 12 by the end of my two terms), I eventually came up with a three-dimensional analytical framework that I refer to as the "driver-policy-people" framework.

Drivers are macro-determinants for the realization, drawbacks, and even violations of human rights. They are indispensable factors in the assessment of how rights can be realized, demonstrating that there are structural barriers that often cannot be addressed at the implementation level. I have elaborated on development cooperation as a driver and an increasing trend in the water and sanitation sector, especially under the SDGs agenda, and my research has identified a significant human rights gap in most of the multilateral or bilateral funders in this regard (reports A/71/302[1] and A/72/127[2]). Another example is megaprojects and the ways that they impact the realization of the rights to water, considering all of the stages in the cycle of megaprojects, from the macro-planning (are they really the only alternative for national development?) to their decommissioning. This issue will be covered in a forthcoming report. Neoliberal policies are other possible drivers, which I am planning to address and elaborate in a future report.

The second dimension, public policies, corresponds to a lower level of determination in the realization of the HRtWS and is more susceptible to the influence of the water sector than other structural roots of the rights realization. Although legislation is fundamental for introducing binding rights in water policies, there are countries that have introduced in their legislation the right to water or the rights to water and sanitation but are not complying with the right's standards. What may induce the alignment of water and sanitation provisions with the HRtWS are policies infused by their framework, with proper accountability mechanisms. There are some elements of public policies that can benefit from this approach. Regulation is one of them, requiring a shift away from traditional economic regulation, which focusses mainly on the financial sustainability of providers, to a regulation framed by human rights (report A/HRC/36/45[3]). Another element sensitive to human rights is financing and the manner in which services balance expenses and revenues, which highlights issues of affordability regarding the access to services as well as the manner in which public finance is used in guaranteeing the access to them (report A/HRC/

30/39[4]). Another example is how the principle of accountability must be introduced in policies, according to the human rights standards (report A/73/162[5]).

The "human" dimension aims to identify the most neglected population groups in relation to access to services, who are neglected either because they bear the greatest burden when services are inexistent or precarious – such as women, girls, or transgender persons – or because they are not targeted by public policies or providers – refugees, migrants, IDPs, homeless, prisoners, etc. (see report A/HRC/33/49[6] on gender issues and A/HRC/39/55[7] on forcibly displaced persons). Giving visibility to those who are invisible might result in different approaches from and in the empowerment of rights holders.

My final observation is that different elements of the framework described previously are addressed in this book and that those elements deliver a two-pronged message. First, that we are increasingly improving our understanding of the human rights to water and sanitation, their tensions, their potentialities, and their fragilities, as a tool for inclusion of those in the most vulnerable situations. Second, that the work is not finalized: we need ever more high-level contributions in this regard. My congratulations and my thanks to Sultana and Loftus, and to the authors of the collection, for their important contributions in filling those gaps.

Léo Heller
UN Special Rapporteur on the Human
Rights to Safe Drinking Water and Sanitation

Notes

1 undocs.org/A/71/302
2 undocs.org/A/72/127
3 https://undocs.org/A/HRC/36/45
4 https://undocs.org/A/HRC/30/39
5 https://undocs.org/A/73/162
6 https://undocs.org/A/HRC/33/49
7 https://undocs.org/A/HRC/39/55

Preface

This book has had an interesting gestation period. When our first edited book – *The right to water: politics, governance, and social struggles* – was published in 2012, it was received well, filling a lacuna in the existing water scholarship on the right to water, while bringing together a large number of international scholars covering a range of interdisciplinary perspectives. The book was translated into two other languages within the first two years of its publication. As a result, we started to receive requests from scholars and students from around the world to produce a second volume. The current book is the product of those requests as well as the persistent inquiries and nudges from our original commissioning editor Tim Hardwick (to whom we are very grateful). But the ultimate decision to bring together another set of excellent scholars was largely inspired by global shifts and events in water-related issues: the debates over the right to water had matured since its formalization in policy in 2010, the manner in which discourses and struggles around the right to water have gone global, and the ways in which scholars, activists, practitioners, and citizens have mobilized around that right in different ways. In addition, the scholarship on the right to water has proliferated in interesting and unexpected ways in recent years. We, therefore, felt it was time to address these issues in another book that could complement our first book and address the recent developments in water politics. This collection broadens existing discussions on the right to water in order to critically shed light on the pathways, pitfalls, prospects, and constraints that exist in achieving global goals, as well as advancing debates around water governance and water justice.

We are grateful to all the authors of the chapters for accepting our invitations to contribute, for being part of this collective journey, for responding graciously to rounds of feedback, and for immeasurable brilliance, camaraderie, and professionalism. A few of the chapters were presented at various conferences and workshops, and we collectively thank feedback from respective participants and discussants. We are grateful to Dr. Léo Heller, United Nations Special Rapporteur on the Human Rights to Safe Drinking Water and Sanitation, for graciously writing the Foreword to the book. For generous endorsements of the book, we thank Professors Erik Swyngedouw (University of Manchester), Quentin Grafton (Australian National University), and Margreet Zwarteveen (IHE Delft). We also thank colleagues at Routledge for seeing the book through to production. Finally, we both are exceedingly grateful to our respective families for supporting us in all our endeavors.

This book, like the first one, is dedicated to all those worldwide who continue to struggle over water and those that fight for justice in water everywhere. The journey continues.

Farhana Sultana and Alex Loftus
Syracuse, NY, and London, UK

1 The right to water in a global context

Challenges and transformations in water politics

Farhana Sultana and Alex Loftus

Introduction

The right to water is widely recognized to have effected a paradigm shift in water governance and water politics. In addition, it has transformed struggles to achieve water justice across scales and sites. In this book, we aim to give some form to such a paradigm shift while simultaneously demonstrating how the right to water has transformed and been translated over the last decade. Much has happened over this time, something we are only too conscious of in reflecting on the years that have passed since we sat down to write the introduction to our last book – *The Right to Water: Politics, Governance, and Social Struggles* (Sultana and Loftus 2012). While published in 2012, we wrote the Introduction in 2011, only one year after the UN General Assembly adopted a resolution recognizing the human right to safe and clean drinking water and sanitation in 2010. In the Foreword to that book, renowned water scholar-activist Maude Barlow, who was then serving as the Senior Advisor on Water to the 63rd President of the UN General Assembly, wrote of the joyous scenes at the UN upon the announcement of the General Assembly's unanimous vote. Our own take was cautiously optimistic. While never ignoring the obstacles and difficulties in achieving the promise of the right to water – in particular, as many others had noted, we were acutely aware of the possibility that the right to water could open the way for a new wave of private sector involvement in the provision of potable water – we were, nevertheless, unwilling to downplay the huge efforts on the part of social movements, non-governmental organizations (NGOs), and scholar-activists who had effectively deployed claims to the right to water as part of a struggle for water justice. Our own perspective was that while the right to water could become an empty signifier, it still represented an important starting point for social mobilizations – a condition of possibility – for broader and deeper struggles for water justice.

Although perhaps something of a cliché to note how much the world has changed since 2010, it is also patently true. The dramatic effects of the 2008 financial crisis were not yet fully clear when we were writing in 2011. And although many may well have accurately predicted that the result would be a growth in inequality, few at the time predicted the viciously revanchist policies now adopted by many governments around the world. At the same time, as the red tide in Latin America stalled – prompted in large part by the collapse of a commodities boom – so several new experiments in social and ecological justice collapsed under the same resource curse that had befallen development projects in the past. Left popular projects appeared to give way to a resurgence of right-wing populism globally, a social project that articulated with a set of economic prescriptions threatening to extinguish of any remaining sparks of socio-ecological justice from the preceding decade. The formal recognition of the right to water was born at a difficult moment in world history. With our cautious optimism, we were thereby forced to confront some brutal realities.

Nevertheless, having witnessed and survived these years, much of what we argued previously seems not far off the mark. The right to water and sanitation remains one of several tools within the armory of those struggling for water justice, whether in the moment of global revanchism we find ourselves now or during the interregnum in which we were writing before. Perhaps surprisingly, this tool now appears to be deployed as readily in the global North as in the global South, despite the lack of formal recognition or adoption of the right in many countries of the former. The movement of struggles and discourses around the right to water from the global South to the global North is, therefore, elaborated upon within this book, in which we now have as many empirical studies of the right to water in the global North as we do in the global South.

Aside from a reconfirmation of our earlier arguments and a broadening of the range of empirical studies, this book nevertheless strikes a different tone. Whereas earlier, the debate among water activists and scholars revolved around whether or not the right to water was likely to be a progressive force in the struggles for water justice, many would now take this progressivism as a given. Instead of debating whether or not the right to water is a positive or a retrograde step, therefore, in what follows, the contributors focus on how best to achieve the right to water in ways that can articulate with other frameworks emerging around water governance, with other conceptions of water justice, and with new perspectives on water security. This book is as much about future trajectories – intersectional and articulating ones – as it is about debating the right to water. In establishing such an argument, the text seeks to provide a set of understandings for new research on water politics more broadly. The arguments put forth here, therefore, collectively animate new possibilities for advancing the right to water. Power constellations and relational understandings of different actors become more evident within these arguments. Indeed, given that the right to water is open to contestation, reinterpretation, and negotiation, multiple potentialities are opened up and alternative imaginaries might be envisioned. Thus, we would reassert the point that the right to water and sanitation remains an important political discourse in supporting the fight for water for the most vulnerable. The right to water remains a profoundly important galvanizing call for pursuing water justice at various scales (see also Sultana 2018; Boelens, Perreault, & Vos 2018).

This book, furthermore, echoes broader shifts – while seeking to capture and analyze them – within its chapters: indeed, the scholarship on the right to water has proliferated in interesting and unexpected ways. Through highly productive sets of conversations, both discourses and struggles around the right to water have: opened new perspectives, politics, and possibilities in water governance; fostered new collective and moral claims for water justice; and effected changes in laws, policies, and institutions around the world. In light of the 2010 UN ratification, changes have taken place in policy, legal frameworks, local implementation, as well as in national dialogues within the majority of countries globally. The novel ways in which the right to water has been taken up in Europe, Latin America, Africa, and Asia point to the enduring appeal and material politics that are enabled and negotiated through this framework in order to address water crises and water insecurities. There is, thus, an urgent need to take stock of debates in light of new concerns around post-neoliberal political developments, the challenges of the Anthropocene and climate change, the transition from the Millennium Development Goals (MDGs) to the Sustainable Development Goals (SDGs), as well as the unexpected mobilizations around the right to water in the global North. This book, therefore, broadens existing scholarship on the right to water globally in order to critically shed light on the pathways, pitfalls, prospects, and constraints that exist in achieving lofty global goals, as well as advancing debates around water governance and water justice.

In this chapter, we frame the book in relation to emerging debates, paying particular attention to intersections with recent discussions in water justice and water governance more broadly. If the right to water is now widely recognized as having forced a paradigm shift in the governance of water and in water politics, it has simultaneously provided a range of tactical and strategic priorities for activists, policymakers, and advocacy groups. Nearly a decade on from the United Nations General Assembly's recognition of the right to water, the present moment provides a unique opportunity for reflecting on the gains, the losses, and the future trajectories for struggles around the right to water. By opening up dialogues with debates around water justice and water governance, we evaluate these gains, losses, and trajectories while also providing a critical framework for new research in the remainder of the chapter.

Institutional questions: whither the state?

Several important institutional questions are important to reflect upon in the present conjuncture. First, as a range of different actors have sought to understand how best to achieve the right to water and in whatever form possible, these debates appear to have recentered questions of the state. How the state is understood, its form and function, as well as its potential capacities have all been questioned. In one of her final contributions to efforts for achieving the right to water, the then UN Special Rapporteur on the Human Right to Safe Drinking Water and Sanitation, Catarina de Albuquerque (2014), developed a Handbook, rather like a toolkit, in which the state is positioned as one of the crucial actors through which citizens can seek to achieve the rights to which they are entitled. In many respects, the centrality afforded to the state is unsurprising, for while the UN may recognize the right to water, it is clearly the member states that are responsible for giving this right any meaning.

Nevertheless, such a position poses some awkward questions. Some of these questions were already noted in our 2012 book. Thus, Bustamante, Crespo, and Walnycki (2012: 223), drawing on the Bolivian experience, had argued that:

> If we are to consider how rights can be recognized and employed by the state, we are recognizing and justifying the state as responsible for ensuring compliance. A rights-based approach means that other institutional and organizational forms are not recognized, even though they may occupy spaces for interaction and rights that don't necessarily originate from the state.

Indeed, although the mechanism that de Albuquerque noted makes intuitive and practical sense, the consequence is a strengthening of the very state institutions that may themselves be responsible for entrenching new forms of hierarchy and perpetuating existing inequalities. Forms of water privatization, to take one example, while appearing to minimize the role of the state, rely on sets of decisions made within the form of the state. Decisions over infrastructure, appropriate levels of investment, and legislation ranging from the ability of water companies to disconnect for nonpayment to the recognition of different groups are all made through the form of the state, very often to the detriment of those who are the least powerful within a given society. Although wrong to generalize from one experience, the South African struggles charted by Clark in the 2012 book, and discussed further in relation to sanitation experiences by Bond in the current book, show how the South African state was able to proscribe a deeper, more participatory, reading of the right to water and, more recently, to inscribe a new "color line" that produces deeply uneven access to different forms of sanitation infrastructure.

Nevertheless, it would be equally wrong to simply dismiss the state as only ever a harbinger of hierarchies or an executive of a racialized bourgeoisie. Indeed, the fact that rates of re-municipalization now outstrip the rate at which water services are being privatized (Kishimoto, Lobina, and Petitjean 2014) – a process that, once again, clearly recenters the state within water provision – is surely something to be celebrated, potentially bringing water services back under some form of democratic control while removing the profit motive with regards to the provision of this most basic of needs. Unsurprisingly, the picture is complicated. As McDonald argues in his chapter in this book, while proponents of privatization are wrong to argue that the involvement of the private sector will bring about the efficiency savings and reduced prices through which the right to water might be realized, the counter-argument that re-municipalization is the only possible way of realizing that right is only partly true. McDonald's chapter, therefore, demonstrates that there is no universal outcome: instead, outcomes are context-dependent, associated with both the form and function of the state, its historical legacies, and the forms of municipal provision that have emerged in relation to that form, function, and legacy.

For some scholars, conversations with work on state theory necessitate a more relational understanding of both the state and of the right to water (Angel and Loftus 2019). Rather than posing a simple question of whether to posit or reject the state as the key agent through which to achieve water justice, it might be possible to adopt an approach that simultaneously works within, against, and beyond the state. Developing such a position necessitates a move beyond more fetishistic understandings that posit the state as a coherent agent capable of enacting particular sets of policies (Abrams 1988). Indeed, it might be possible to think of strategies that move within, against, and beyond the right to water.

Whereas Clark (2012), Angel and Loftus (2019), and the chapters by Bond and McDonald all tend to focus predominantly on the national and local state, the chapters by Bieler and by Van den Berge et al. demonstrate the ways in which supranational institutions, such as the European Union (EU), have also come to mediate particular struggles around the right to water. In a remarkable feat of mobilization, water activists were able to achieve the first-ever Citizens' Initiative within the European Union, thereby paving the way for a debate over the right to water within the European Commission. The Citizens' Initiative, coordinated by a broad coalition of trade unions and civil society organizations, gathered over 1.9 million signatures across 14 of the 27 countries comprising the EU, in order for the right to water to be prioritized within EU legislation. While certainly not an unqualified success, the Citizens' Initiative serves to demonstrate the ways in which new institutional frameworks now embody and express the different struggles for water justice and, furthermore, how new tools are opened up at different scales – and fundamentally different locations – for pursuing greater water equity.

The chapter by Mehta and Langmeier, alongside that by Schmidt and by Meehan, show the ways in which the UN, through its High-Level Expert Panel has also sought to further clarify the roles and responsibilities of different institutions within a broader governance framework. Schmidt charts in forensic detail the evolution of discourses within the UN High-Level Panel on Water and the likely implications of apparent shifts in focus. By beginning to integrate questions around resilience – in its turn to "Valuing Water" – the arguments in favor of non-contingent human rights appear to lose their force. Instead, and problematically, explanations of water scarcity appear to be rooted more in instances of "moral luck." Such discourses clearly matter both for the prominence or otherwise of the right to water and for the political pressure that can be exerted on institutional frameworks in ensuring equitable access to water. As charted in the chapter by Mehta and Langmeier, Mehta's own involvement in the High-Level Panel of Experts on Food Security and Nutrition enabled her to push for far greater attention to the necessary relationship between the right to food and to water, even against the apparent wishes of the Special Rapporteur at the time, for whom such a focusing would distract from the more immediate task of achieving change in the water sector. The Foreword to this book from the current Special Rapporteur, Léo Heller, furthermore, demonstrates the continued importance of the Special Rapporteur's role and for the position to be able to shape and influence debates, both in terms of the particular lens brought to the analysis and through the thematic reports covered in their role. These institutional levels, positions, platforms, and networks clearly matter.

Water security discourses

If institutional structures have shifted and new institutional forms have come to mediate struggles over the right to water, a range of discourses have also emerged that may or may not be complementary to struggles to achieve that right. Perhaps most prominent among these is a whole set of discussions over how best to achieve water security. Such discussions then pose questions around the degree to which efforts to make communities "water secure" can be considered to be the same as affording the right to water.

The turn to water security in recent decades should be distinguished from earlier discussions – emerging from realist approaches in International Relations (IR), theory – in which water security's primary referent object was the nation-state. In the immediate post-Cold War era, discussions over environmental security appeared to be enrolled in efforts to script the new threats faced by the world. Often couched in thinly disguised – or even avowedly – neo-Malthusian frames, water security was said to pose dangers to national security and to require both careful military planning and new engineering solutions (Starr 1991). Critical perspectives remained unsurprisingly wary of such concerns, and instead sought to demonstrate the ways in which water insecurity is socially produced in relation to broader classed, sexed, racialized, and gendered social relations. Discussions over environmental security have transformed in recent decades, enabled by a shift to a human security framing (in which the referent object shifts from the nation-state to the individual and community) (Barnett 2001; O'Brien 2006) and, for others, to new sets of questions around ecological security, in which the biosphere becomes the principal referent object to be secured (Detraz 2009; Cudworth & Hobden 2011).

But if the referent object has shifted, discussions of water security do not always manage to avoid the Malthusianisms or environmental determinisms of previous iterations. Thus, for Grey and Sadoff (2007) water insecurity can be read as affecting those who are "hostages to hydrology." Both the language and the framing clearly view the environment as shaping the fate of individuals – one's exposure to environmental harms is seen to result from the quirk of one's birthplace rather than the sets of relations that give rise to particular forms of inequality. It is against such dominant framings that critical approaches to water security have emerged, and it is these critical approaches that might influence the future direction of critical approaches to the right to water. Within this book, the chapter by Jepson et al. employs a capabilities approach in order to argue that water security is more about securing the relationships through which water shapes people's lives. Water security should, therefore, be understood less in terms of securing water as an object. Instead, critical approaches to water security should build on previous relational understandings, thereby enabling new synergies to emerge between the right to water and water security (see also Loftus 2015).

Mehta and Langmeier's chapter similarly emerges from a set of debates over the right to food security and nutrition. As with discussions over water security, food security has come to be framed in fundamentally different ways from the Cold War security discourses of realist IR. Interlinkages are now emphasized such as those teased out carefully by Mehta and Langmeier with the right to water: the referents are no longer nation-states but people, communities, and human subjects. Discussions of ecological security – often emerging through post-human approaches to IR (Cudworth and Hobden 2011) – have sought to shift the discussion more towards a consideration of the security of both human and nonhuman, suggesting a need to consider the rights of the nonhuman alongside the rights of the human. Although we approach some of these discussions with slight skepticism, schooled as we are in critical political ecological debates that reject claims to "the natural," as we discuss in the next section, critiques of the knowledge base of settler colonialism have developed a political ontology that unsettles some of these certainties in important ways.

Articulating race/class/indigeneity/coloniality

As with the relational approaches developed around critical approaches to water security, recent discussions have also emphasized the ways in which water injustice always exceeds the socioeconomic inequalities through which it is, in part, produced. Indeed, the social productions of race, class, and gender articulate with and are reinforced through access to water. The right to water needs to be understood within such a framework, not only because it has implications for how we understand the articulations of social difference and access to water, but also because social difference and access to water have material and reciprocal effects.

To provide some context, scholars have shown how water is intersectionally gendered, racialized, and classed in different contexts (e.g. Brown 2010; Harris 2009; O'Reilly 2006; O'Reilly et al. 2009; Sultana 2009, 2011). In so doing they have elucidated the impossibilities of single narratives on water in any location. Hardships and struggles have to be analyzed in nuanced ways to underscore the heterogeneity of power relations involved. Scholars focusing on the global South have provided a plethora of analyses that can further articulate with the right to water, from a range of perspectives, such as postcolonial, feminist, critical race, and critical urban studies (e.g. Gandy 2008; Harris et al. 2017; Hellum, Kameri-Mbote, & van Koppen 2015; Kooy 2014; Sultana 2009; Sultana, Mohanty, & Miraglia 2016). In any context, water insecurity, unavailability, and stress will exacerbate existing intersectionalities of power differences. Investigating the ways water marginalization follows the contours of historical and intersectional dispossessions fosters a greater understanding of inequities on the ground. For instance, the chapter by Clark shows how racialized structural violence is perpetuated through water injustices in Flint, Michigan, USA. The (re)production of racialized difference within post-apartheid South Africa is also captured in the chapter by Bond, focusing on sanitation struggles in Durban. And the chapter by Zenner shows how a focus on indigeneity raises ontological questions and challenges in materializing a right to water.

If the continuities and inheritances of colonial systems of inequality have been widely acknowledged within the literature on the global South – thereby emphasising how such relations have shaped the infrastructure networks, as well as the systems of governance, through which water is provided in different parts of the postcolonial world – more recently, attention has also turned to questions of settler colonialism as a specific set of social relations that shapes the colonial present of North America, Australia, New Zealand, and Palestine. Ontological certainties over the hydrologic cycle, the right to water itself, and the modes through which it might be realized have been called into question as the assumptions of settler colonial states, along with Eurocentric frameworks, have been called into question. Human rights frameworks might, therefore, be read as a Eurocentric application of a liberal framework that guarantees the entitlements and freedoms of some over others. The need to disrupt – or at least question – settler colonial understandings is perhaps best seen in the chapter by Linton in which he poses the question of how the right to water sits with the settler colonial problematic. For Linton, troubling settler colonial relations forces the question of how the right to water might adequately account for fundamentally different ontological framings of water: water-as-lifeblood is a rights-producing ontology. It is the right to perform the enactments and practices associated with water-as-lifeblood and respecting such a right and enabling its achievement requires recognizing the right to ontological difference. Echoing the work of Yates, Harris, and Wilson (2017), Linton demonstrates how right to water discourses might be brought into conversation with other non-European ontologies and non-anthropocentric worldviews. The challenges posed in recognizing the right to ontological difference are many and, as Linton demonstrates, not easily or comfortably resolved. Zenner, in a slightly different vein, argues in her chapter that different ontologies are incommensurate with – and sometimes even oppositional to – more anthropocentric understandings of the right to water. She teases out such an analysis through her engagements with two indigenous movements – the *mni wiconi* at Standing Rock, USA, and the conferral of legal personhood to the Whanganui River in Aotearoa, New Zealand – alongside the moral proclamations on the right to water by the Catholic Church. Together, these chapters bring into sharper focus the need for nuanced reflexivity and an understanding of multiple framings and ontologies in any discussion of the right to water.

Translating the right to water from the South to the North

As we noted at the outset, one of the differences from our earlier book is the number of studies within this new book that draw on experiences from the global North. The right to water has become a crucial tool for water justice activists in the global North in recent years seeking to achieve fairer access to water, contesting different forms of privatization and commercialization, and exposing and challenging the effects of racial capitalism as well as settler colonialism. Each of the chapters focusing on the global North shows the emergent and existing ways that the right to water has been deployed as discourse, practice, imaginary, solidarity, and critique. Contextually-rich, these chapters show how such discourses and practices work within spaces of imperialism, through-and-against the relations of advanced capitalism, and through the infrastructural forms of so-called advanced industrialized countries. Initially, the right to water was expected to galvanize further social movements in the global South, whereas now the right to water has truly gone global. Concerns around poverty, inequalities, and dispossessions provide challenges to fulfilling livelihood needs globally, mobilizing different groups to make moral and material claims around water.

The challenges of deploying the right to water within the racial capitalism of the global North is seen within the chapter by Clark, focusing on the case of Flint, Michigan. Whereas Clark, drawing on Bello (2004), makes the claim that Flint is as much part of the global South as the global North, elsewhere Ranganathan (2018) urges against Third World comparisons in the case of Flint, suggesting that these obviate the specific ways in which racial capitalism has developed and unfolded within the United States. Clark is in no way blind to the "racial liberalism" through which Ranganathan (2016) structures her analysis of Flint, but through her own critical legal scholarship, Clark is able to draw connections between her earlier work in Johannesburg and the struggles for water justice in Michigan. In so doing, Clark's chapter demonstrates how struggles that largely emerged from the global South have now come to profoundly shape efforts for water justice in the global North, including in contexts where the right to water has not been formally adopted by nation-states.

Given Clark's attention to both contextual specificity and the possibility for shared struggles to develop across both North and South, perhaps it might also be possible to read Flint – and the right to water more broadly – through a form of relational comparison (Hart 2006, 2018). Hart's understanding of relational comparison is perhaps best seen in her rich explorations of the divergent trajectories of Ladysmith and Newcastle in South Africa, always understood in their relations to global processes that link land tenure in East Asia to forms of industrialization and racialized dispossession in South Africa. For Kipfer and Hart (2013: 323) relational comparison requires paying attention "to the spatio-historical, articulatory, and denaturalizing aspects of translating practice." In relation to the right to water, we might, therefore, begin to consider the manner in which the political practice of achieving the right to water articulates with and through the contextual specificities (always relationally understood) of specific sites.

In this regard, the practicalities of the right to water must always be seen as simultaneously contextual, discursive, and material albeit always articulating with broader globalizing relations. Thus, Walnycki's chapter on Bolivia shows us the complexities involved in state-community relationships to actualize what the right to water could mean. She argues that there has been a gradual de-politicization of the global campaign to realize the human right to water as it has been subsumed under Sustainable Development Goal 6 (water). The practical and strategic innovations required to implement the right to water in complex under-served urban areas in the global South is found to be lagging even if numerous informal providers can enable the progressive realization of the right to water. Non-state actors are often essential in materializing the right to water, especially from the informal sector, NGOs, and grassroots mobilizations (see also Wutich, Beresford, and Carvajal 2016). In another chapter, Meehan shows how different institutional actors operate at a range of different scales in Mexico in order to develop practical authority for human rights through a diverse set of sites, tactics, and strategies. Moving beyond constitutional scripts and policy impasses, Meehan argues for the necessity of attending to the different and alternative pathways by which the right to water is materialized on the ground. Both Meehan's chapter and Walnycki's chapter are reminders of the continued struggles and battles being fought in spaces where formal adoption of the right to water exists but where there are no straightforward solutions. They also offer examples of the contextual specificities and relationalities fostering and hindering the achievement of the right to water.

While perhaps more prominent as a theme in our earlier book, encroaching commoditization of water is often identified as one of the key shifts enacted through capitalist relations of production within the hydrosocial cycle. One of the crucial touchstones within the previous book was Bakker's argument that commoning strategies might provide a more useful strategy for countering commoditization than struggles for the right to water, given the aforementioned risk that the right to water is used as an opening for new forms of private sector involvement in water delivery (Bakker 2007). Given the trend to re-municipalize water, debates over commoditization are somewhat less present within this current book – although see the chapters by van den Berge et al. and by Bieler for important and thoughtful takes on such questions. A somewhat trickier set of questions emerges in relation to the massive growth of the bottled water industry, as highlighted in the chapter by Pacheco-Vega, in which he shows how the right to water can be instrumentalized by the bottled water industry, yet at the same time packaged water can foster the fulfillment of the right to water in times of disasters and crises. Developing a comparative analysis of two disasters, one in the United States and one in Mexico, his chapter demonstrates the ways that bottled water sits uncomfortably within the majority of the right to water discourses. Disasters create immediate demands for a greater supply of packaged/bottled water to distressed communities. At the same time, the rise of bottled water consumption, when easily accessible safe potable water exists via public provisions, continues to raise questions of intensifications over commodification and privatization of water sources (see also Hawkins 2017).

While we would concur that the capture of human rights discourses by market forces can undermine the realization of community water rights as well as a notion of the commons in water, we simultaneously wish to show how contingent relations, contours, and struggles over water open up spaces for greater democratic, participatory, and equitable interventions and possibilities. None of this is monolithic nor utopian. Forms of political praxis aimed at realizing the right to water have always been birthed through struggles. Critical attention must, thus, remain focused on issues of elite capture, participatory exclusions, and marginalizations across intersectionalities of gender, race, class, and other axes of differences. The chapters in this book, therefore, deconstruct public/private binaries, emphasizing the complex terrains within a maturing set of understandings over the right to water, and bring into focus the different actors, options, processes, and challenges involved. Issues of accountability, decision-making, cooperation, and transparency are shown to be central in realizing a right to water, since normative claims, devoid of politics and materiality, do not necessarily result in actualized gains.

Conclusion

In looking backward and thinking forwards while pondering a decade of water rights and wrongs, this book discusses the multiplicity of possibilities around how to achieve the right to water rather than only theorize what it is or could be. The book is more about the struggles and strategies for thinking through the right to water and the tensions therein. The chapters collectively encourage learning from the different frameworks in relation to that right, the various struggles that have worked with and through the right to water, and those that have reinterpreted what it means on the ground. What we see now are various power constellations and relational understandings from different actors involved in ensuring the right to water in different contexts. Access to safe, clean water by differently situated groups is a product not necessarily of one independent actor but usually several in constellations of power.

While legal strategies and policy instruments are often thought to be the common outcomes of adoption of rights discourses by nation-states, it is certainly not limited to these, as the right to water has been found to offer strategic tools for organizing, mobilizing, and resisting across scales and locations. The right to water is not just about quantity, quality, availability, or access but fundamentally about the right to participate in water governance and power structures that influence those rights. Broader goals of justice and equity are being envisioned and pursued globally at the current conjuncture. Emancipatory politics in the Anthropocene is being made more possible through the different ways that the right to water has been taken up globally. This has important ramifications for water governance and water justice in practice and in discourse. The right to water has enduring appeal precisely because it offers ways to move past reductionist prescriptions that render technical the things that are very much political. The malleability of the term offers a politics that can engage anticolonial, decolonial, anti-race, feminist, and other intersectional strategies to contextualize grounded realities. The right to water has become a galvanizing call to reimagine and co-construct a politics that works from the bottom up and holds multiple possibilities for hope and social justice in relation to water.

References

Abrams, P. (1988) 'Notes on the difficulty of studying the state,' *Journal of Historical Sociology*, 1 (1), pp. 58–89.

Angel, J. and Loftus, A. (2019) 'With-against-and-beyond the human right to water,' *Geoforum*, 98, pp. 206–213.

Bakker, K. (2007) 'The "commons" versus the "commodity": alter-globalization, anti-privatization and the human right to water in the global south,' *Antipode*, 39 (3), pp. 430–455.

Barnett, J. (2001) *The meaning of environmental security: ecological politics and policy in the new security era.* London: Zed Books.

Bello, W. (2004) 'The Global South,' in Mertes, T. (ed) *A movement of movements: is another world really possible?* London: Verso.

Boelens, R., Perreault, T., and Vos, J. (eds.) (2018) *Water justice.* Cambridge: Cambridge University Press.

Brown, R. (2010) 'Unequal burden: water privatisation and women's human rights in Tanzania,' *Gender and Development*, 18 (1), pp. 59–67.

Bustamante, R., Crespo, C., and Walnycki, A.M. (2012) 'Seeing through the concept of water as a human right in Bolivia,' in Sultana, F. and Loftus, A. (eds), *The right to water: politics, governance, and social struggles.* Abingdon: Routledge, pp. 223–240.

Clark, C. (2012) 'The centrality of community participation to the realization of the right to water: the illustrative case of South Africa,' in Sultana, F. and Loftus, A. (eds), *The right to water: politics, governance, and social struggles.* Abingdon: Routledge, pp. 174–189.

Cudworth, E. and Hobden, S. (2011) 'Beyond environmental security: complex systems, multiple inequalities and environmental risks,' *Environmental Politics*, 20 (1), pp. 42–59.

de Albuquerque, C. (2014) *Realising the human rights to water and sanitation: a handbook.* Available at: http://unhabitat.org/series/realizing-the-human-rights-to-water-and-sanitation

Detraz, N. (2009) 'Environmental security and gender: necessary shifts in an evolving debate,' *Security Studies*, 18 (2), pp. 345–369.

Gandy, M. (2008) 'Landscapes of disaster: water, modernity, and urban fragmentation in Mumbai,' *Environment and Planning A: Economy and Space*, 40 (1), pp. 108–130.

Grey, D. and Sadoff, C.W. (2007) 'Sink or swim? Water security for growth and development,' *Water Policy*, 9 (6), pp. 545–571.

Harris, L. (2009) 'Gender and emergent water governance: comparative overview of neoliberalized natures and gender dimensions of privatization, devolution, and marketization,' *Gender, Place, and Culture*, 16 (4), pp. 387–408.

Harris, L., Kleiber, D., Goldin, J., Darkwah, A., and Morinville, C. (2017) 'Intersections of gender and water: comparative approaches to everyday gendered negotiations of water access in underserved areas of Accra, Ghana, and Cape Town, South Africa,' *Journal of Gender Studies*, 26 (5), pp. 561–582.

Hart, G. (2006) 'Denaturalizing dispossession: Critical ethnography in the age of resurgent imperialism,' *Antipode*, 38 (5), pp. 977–1004.

Hart, G. (2018) 'Relational comparison revisited: Marxist postcolonial geographies in practice,' *Progress in Human Geography*, 42 (3), pp. 371–394.

Hawkins, G. (2017) 'The impacts of bottled water: an analysis of bottled water markets and their interactions with tap water provision,' *Wiley Interdisciplinary Reviews: Water*, 4 (3), pp. 1–10.

Hellum, A., Kameri-Mbote, P., and van Koppen, B. (eds) (2015) *Water is life: women's human rights in national and local water governance in Southern and Eastern Africa*. Harare: Weaver Press.

Kipfer, S. and Hart, G. (2013) 'Translating Gramsci in the current conjuncture,' in Ekers, M., Hart, G., Kipfer, S., and Loftus, A. (eds) *Gramsci: space, nature, politics*. Oxford: Wiley.

Kishimoto, S., Lobina, E., and Petitjean, O. (2014) *Our public water future: The global experience with remunicipalization*. Transnational Institute (TNI)/Public Services International Research Unit (PSIRU)/Multinationals Observatory/Municipal Services Project (MSP)/European Federation of Public Service Unions (EPSU). Available at: www.tni.org/files/download/ourpublicwaterfuture-1.pdf

Kooy, M. (2014) 'Developing informality: the production of Jakarta's urban waterscape,' *Water Alternatives*, 7 (1), pp. 35–53.

Loftus, A. (2015) 'Water (in) security: securing the right to water,' *The Geographical Journal*, 181 (4), pp. 350–356.

O'Brien, K. (2006) 'Are we missing the point? Global environmental change as an issue of human security,' *Global Environmental Change*, 16, pp. 1–3.

O'Reilly, K. (2006) '"Traditional" women, "modern" water: linking gender and commodification in Rajasthan, India,' *Geoforum*, 37 (6), pp. 958–972.

O'Reilly, K., Halvorson, S., Sultana, F., and Laurie, N. (2009) 'Introduction: global perspectives on gender-water geographies,' *Gender, Place, and Culture*, 16 (4), pp. 381–385.

Ranganathan, M. (2016) 'Thinking with flint: racial liberalism and the roots of an American water tragedy,' *Capitalism Nature Socialism*, 27 (3), pp. 17–33.

Ranganathan, M. (2018) 'Beyond "third world" comparisons: America's geography of water, race, and poverty,' *International Journal of Urban and Regional Research*, Spotlight Series. Available at: www.ijurr.org/spotlight-on/parched-cities-parched-citizens/beyond-third-world-comparisons-americas-geography-of-water-race-and-poverty

Starr, J.R. (1991) 'Water wars,' *Foreign Policy*, 82, pp. 17–36.

Sultana, F. (2009) 'Fluid lives: subjectivities, gender, and water in rural Bangladesh,' *Gender, Place, and Culture*, 16 (4), pp. 427–444.

Sultana, F. (2011) 'Suffering for water, suffering from water: emotional geographies of resource access, control, and conflict,' *Geoforum*, 42 (2), pp. 163–172.

Sultana, F. (2018) 'Water justice: why it matters and how to achieve it,' *Water International*, 43 (4), pp. 483–493.

Sultana, F. and Loftus, A. (eds) (2012) *The right to water: politics, governance, and social struggles*. Abingdon: Routledge.

Sultana, F., Mohanty, C.T., and Miraglia, S. (2016) 'Gender equity, citizenship, and public water in Bangladesh,' in McDonald, D. (ed) *Making public in a privatized world*. London: Zed Books, pp. 149–164.

Wutich, A., Beresford, M., and Carvajal, C. (2016) 'Can informal water vendors deliver on the promise of a human right to water? Results from Cochabamba, Bolivia,' *World Development*, 79, pp. 14–24.

Yates, J.S., Harris, L.M., and Wilson, N.J. (2017) Multiple ontologies of water: politics, conflict, and implications for governance. *Environment and Planning D: Society and Space*, 35 (5), pp. 797–815.

2 Valuing water
Rights, resilience, and the UN High-Level Panel on Water

Jeremy J. Schmidt

Introduction

Human rights are designed to be non-contingent; equality independent of moral luck. Geographical location, gender, ethnicity, or religion are just a few markers that human rights are intended to operate free from. In practice, however, human rights have been "kindest to the rich," as the pursuit of equality has been displaced by norms of sufficiency—largely couched in terms of human needs—that are unconcerned with inequality so long as subsistence minimums are met (Moyn, 2018). For instance, the 1948 Universal Declaration of Human Rights begins with the recognition that "the inherent dignity" and "equal and inalienable rights of all members of the human family is the foundation of freedom, justice and peace in the world" (United Nations General Assembly, 1948). Yet, when the Human Right to Water and Sanitation (HRWS) was under consideration in 2010, the expert report to the United Nations rarely referenced equality. Instead, experts stressed norms of sufficiency, such as in statements like, "The normative content of the rights to water and sanitation can be determined in terms of the criteria of availability (referring to sufficient water for personal and domestic use, or sufficient sanitation facilities)" or assertions that "Water supply for each person must be sufficient for personal and domestic uses" (United Nations General Assembly, 2010, p. 10, 11). In this chapter, I argue sufficiency approaches to the HRWS naturalize moral luck at the expense of equality in programs of global water governance. I show how by examining the "valuing water" initiative of the United Nations High-Level Panel on Water (HLPW), which seeks to unite the HRWS, the Sustainable Development Goals, and the profound impacts of humans on the Earth system.

My argument builds on two earlier interventions. In the first, I outlined an approach to ethics and the HRWS grounded in an appreciation of the judgments through which states of affairs are rendered into governance propositions, such as water scarcity or water security (Schmidt, 2012). In the second, I examined how the international uptake of such propositions has naturalized global hydrology to the management institutions and governance norms of political liberalism (Schmidt, 2017). Those accounts rely, in part, on Hannah Arendt's (1958a, 1982) arguments regarding how shared practices produce common sense judgments—a *sensus communis*—and her insights into the politics of shifting from fixed categories of "nature" to sciences that disclose how humans have channeled nonhuman processes (like nuclear energy) into social worlds. In this chapter, I use both ideas to draw attention to Arendt's provocative notion of a "right to have rights." I do so in the context of how the HLPW has sought new water values in order to align sufficiency-based approaches to the HRWS with scientific accounts of how human, economic processes are integrated with planetary ones. As Arendt (1958b: p296) wrote:

> We became aware of the existence of a right to have rights (and that means to live in a framework where one is judged by one's actions and opinions) and a right to belong to some kind of organized community, only when millions of people emerged who had lost and could not regain these rights because of the new global political situation.

Arendt's idea of a "right to have rights" was developed in the aftermath of WWII when millions of people were rendered stateless. But it can also help to make sense of how global water governance programs square non-contingent human rights with claims that the resilience of the Earth system is being profoundly impacted by human forces. To explain how this is the case and why it matters to the HRWS, the chapter is divided into three parts. The first expands further on a non-contingent notion of human rights to sharpen the distinction between equality and sufficiency. The second examines the naturalization of moral luck through resilience. The appeal to resilience does not reject rights; instead, it disarms the political notions of community that are central to a "right to have rights" and which are crucial to equality. The third section examines how the HLPW used its 24-month "valuing water" initiative from 2016–2018 to connect resilience, the HRWS, and the 2015 Sustainable Development Goals. The HLPW launched in 2016 as a joint endeavor of the United Nations and the World Bank. It is comprized of 12 current or former heads of state and began its work by arguing that "valuing our water right" requires "building resilience to climate change" and "allocating [water] to areas of highest social, economic, and environmental value, whether through policy or pricing mechanisms" (UN HLPW, 2016, p. 2). The embrace of human rights and both economic and environmental resilience, however, also cements sufficiency over equality at the expense of a satisfactory account of the political community. The chapter concludes by arguing that water is a prerequisite to belonging to a political community, itself a requisite for a "right to have rights."

The non-contingent nature of human rights

What does it mean to say human rights are non-contingent? It does not mean they are universal or timeless. Human rights quite obviously depend on historically specific ideas, actors, and institutions. Namely, those that led from the Universal Declaration of Human Rights to, five decades later, an extension to water and sanitation; other works in this book and beyond position the sited contests over theory, policy, and practice (e.g. Langford and Russell, 2017). Human rights are non-contingent in the sense that the moral luck of one's station in life does not affect whether or not an individual qualifies for such rights. Henry Shue (1980) argued that basic rights, such as to the security of one's person and to minimal subsistence (e.g. clean air, an unpolluted environment), are prerequisite to other kinds of rights. Basic rights issue from moral claims regarding individual goods and are justified by what individuals may reasonably demand with respect to those goods. In normative theory, a distinction is often made between positive and negative rights (although Shue rejects this distinction). The latter is shaped by principles of noninterference: the right to not have individual freedoms constrained by others or by the state and to act in a way that does not interfere with others. Positive rights, by contrast, are based on entitlements to some type of good, such as protection by the state without discrimination, inclusion in social welfare programs, or access to water.

There are good reasons to be skeptical about a hard division between positive and negative rights when it comes to water. For instance, negative rights depend on recognition of individuals or groups who hold rights that oughtn't be interfered with. For the vast majority of the world's population, however, such political recognition is not a given, but a hard-won status. In these cases, it is critical for individuals and communities that the state recognize them in ways that qualify them to be governed as citizens who hold rights and entitlements to material and moral goods (Chatterjee, 2006). In Mumbai, as Nikhil Anand (2017) shows, the politics of the governed plays out in complicated contests as people negotiate, strategize, and elbow their way into being recognized as citizens owed water services by applying various kinds of material and moral pressure—they both demand to be governed on moral grounds and take up material opportunities to have water demands met. Such situations do not fit neat distinctions between positive versus negative rights because to be governed is to be interfered with—at least to some extent—and in the case of water that interference is central to satisfying a basic right to water that (once secured) should not itself be interfered with.

Consider another concern. Negative rights require spaces where actions by one person do not interfere with the rights of another. Formerly, international water management could rely on a natural framework for identifying the conditions of noninterference in reference to stationarity: "the idea that natural systems fluctuate within an unchanging envelope of variability" (Milly et al., 2008, p. 573). The hydrological cycle's natural envelope of variability implied that, while there may be fluctuations year-to-year, probability distributions could be calculated in ways that made rational planning possible and which would keep interference among rights holders to a minimum, typically states sharing transboundary waters in the case of international water management (see Schmidt, 2017). But stationarity can no longer be assumed owing to how human actions have pushed the global climate system into a novel state that has changed the envelope of hydrological variability itself (see Milly et al., 2008). By altering the outer bounds of variability, anthropogenic climate change upends the idea that there is empty space left for the Earth to absorb negative externalities—the actions of one actor that may interfere in harmful ways with another even if unintentionally—as they affect the water cycle. On a full Earth, every unit of carbon dioxide emitted interferes not only with the water cycle but with those holding or seeking water rights. Not only does this trouble the empirical prospects for the principle of noninterference but it also frustrates efforts to determine whose emissions infringe on whose rights and amplifies considerations of climate and water justice (see Jamieson, 2014; Shue, 2014; Boelens, Perreault, & Vos 2018).

Morally, human rights are not circumscribed by circumstance. Often, however, the rational defense of human rights has attempted to treat them as placeless. This feature of human rights is frequently critiqued for issuing from western, liberal conceptions of the human subject. For example, claiming that humans share universal traits, such as autonomy or rationality, and then using these traits to defend human rights may be critiqued for not recognizing social or cultural difference. The initial draft of the Universal Declaration of Human Rights, it can be recalled, was penned by the Canadian John Humphrey, who was deeply influenced by the French philosopher Henri Bergson and his arguments about the vitalism inherent to all evolutionary beings (Curle, 2007). Clearly, there are good reasons to be skeptical of universal or naturalized defenses of human rights. But naturalizing the reasons for defending human rights is only part of what might be questioned. Consider, for instance, the argument of Karen Bakker (2012, p. 37, emphasis in original) who holds that, in meeting the HRWS, "focus on *property rights* rather than human rights offers us more potential for politically progressive strategies." Bakker's claim considers naturalism from a different direction, from claims about individual traits to those of social institutions, such as property. There are good reasons to be skeptical about emphasizing property rights rather than human rights, since this makes basic rights subordinate to the political economy of social institutions (see Schmidt and Mitchell, 2014). For instance, in settler-colonial states such as the United States and Canada, the social institutions of public and private property rights are premised on Indigenous dispossession (Hall, 2010; Byrd, 2011; Bhandar 2018). Further, numerous legal scholars argue property rights are also biased towards liberal conceptions of the relationships among individuals and society and, in addition, are environmentally pathological, owing to how they carve up ecological systems in piecemeal, instrumental fashion (Butler, 2000; Boyd, 2017).

If we accept the critique that human rights claim to be non-contingent but that justifications are often culturally specific, does this undermine the HRWS? We might be tempted to believe so until the many ways that the HRWS is claimed, valued, and positioned by different groups come into view (e.g. Hellum et al., 2015; Perera, 2015). Many strategic appeals to human rights refuse norms of sufficiency and instead make human rights instrumental to the pursuit of equality. In this regard, human rights can be used to address social and economic structures, such as private property, if and when they produce inequality.

As Moyn (2018) argues, the rise of human rights in the late 20th century paralleled the rise of social and economic inequality. In the water sector, the replacement of rights-based approaches to equality with needs-based approaches to sufficiency underpin Eckstein's (2010) arguments for flexible approaches to water security in view of both climate change and the potential mismatch between legal claims and water availability. But equality does not mean material equivalence. Rather, equality implies that human rights bend social, economic, and legal structures evermore towards justice. In this way, human rights can work in ways that do not depend on agreement with a particular view of the "subject" who holds rights but rather on manipulating that subject (even if a legal fiction) into a position where claims can be made that might be otherwise unavailable to different individuals or groups. In this sense, human rights may provide a mechanism for countering the oppression of existing rights regimes, such as those of private property, when they are used to dispossess others of rights to land and water. Yet, while use and extension of human rights to counter oppression have a satisfying ring, the pursuit of equality is often overshadowed by sufficiency-based norms of global water governance.

The contingent world of resilience

In 2014, in response to humanity's massive disturbance to the global water system, hydrologists called for a shift towards a resilience-based approach to water governance (Rockström et al., 2014). It was not the first such call. Experts have long argued that adaptive management is best suited for governing water in social-ecological systems to enhance resilience, meaning the capacity of a system to withstand disturbances and still maintain its characteristic identity, functions, and feedbacks (Folke, 2003; Falkenmark and Rockström, 2004; Feldman, 2007). For many, resilience promises flexibility in the pursuit of sustainable development, such as in integrated water resources management (IWRM), because it recognizes change, not stability, characterizes complex social, economic, and environmental systems (Galaz, 2007; Pahl-Wostl, 2015). However, resilience-based approaches also carry distinct political implications for the HRWS owing to how moral luck is naturalized in a way that disarms political notions of community.

Resilience emerged in the 1970s as a way to understand ecological relationships and has since expanded to consider the dynamics of complex, adaptive dynamics of coupled social-ecological systems (see Folke, 2006). As the concept matured, it moved from a metaphor for explaining system dynamics to a way of measuring social-ecological capacity (Carpenter et al., 2001). Ultimately, it gained global purchase owing, in part, to the worldwide network of scientists who positioned resilience as central to the Millennium Ecosystem Assessment, which published its results in 2005. Over the past decade, resilience has gained a virtual hegemony in programs seeking to link water to other domains of sustainable development, such as the water-energy-food nexus that is increasingly central to the Sustainable Development Goals (e.g. Biggs et al., 2015). Resilience has also ascended owing to its use by international financial networks, such as the World Economic Forum, that have promoted resilience as the way to link concerns of governance and security with interconnected risks and uncertainties to the Earth system, the water-energy-food nexus, and the global economy (Schmidt and Matthews, 2018).

The rise of resilience was not only an outcome of scientific networks but also of its uptake in economics and sustainability. In this regard, resilience fit with the turn to complexity science in economics that also emerged in the 1970s and which took on special significance after the 2008 global financial crisis (Cooper, 2010, 2011). In particular, explanations of financial crises appealed to ecological theories of complexity to frame and address intersecting economic and environmental crises in terms of resilience (Walker and Cooper, 2011). And it was not only economists and financial planners who now advance this perspective. Ecologists and political theorists argue there is a shared "causal architecture" linking crises across the water-energy-food nexus, the Earth system, and the global economy (see Homer-Dixon et al., 2015). Similarly, the Global Water Partnership (2012) began realigning IWRM from the traditional sustainable development definitions that it had helped to solidify through the 1990s—coordinating water development and management to maximize social and economic welfare without compromising vital ecosystems—to a resilience-based approach to water security that includes the enhanced use of financial instruments to address coalescing challenges of environmental and economic sustainability.

As resilience ascended as a way to link environment and development, it was also installed as a way to link new understandings of the planet to global sustainability agendas. As it did, a tension arose between non-contingent human rights and empirical facts that the Earth system was contingent. Like the global water cycle, the way the Earth system functioned did not vary within an unchanging envelope but was, in fact, being pushed into a "no analogue" state by human forces (Steffen et al., 2004). In fact, the findings of Earth system science prompted a normative shift away from seeking to integrate environment and economy within planetary limits—the original impetus for sustainable development (see Macekura, 2015)—and towards recognition of how humans and the Earth were already integrated. Subsequently, human-caused changes to the Earth system were linked to sustainability agendas through the notion of "planetary boundaries" for freshwater use, climate change, biodiversity integrity, and six other domains. If planetary boundaries are not transgressed, they provide a safe space for development and a framework for understanding the links of development, finance, and sustainability (Sachs, 2015). Recent assessments of the Earth system, however, suggest it may be on a trajectory to cross critical thresholds that close off paths of return to a more stable, resilient planet (Steffen et al., 2018). Accelerating pressure on the planet has, in turn, led to calls for a "green economy" that many in the Global South worry will lead to new forms of financial discipline that shift sustainability norms away from the pursuit of equality and justice (see Conca, 2015). All told, the gathering momentum towards bundling environmental and economic crises together with resilience-based approaches to sustainability has shifted the integration of humans and the planet from a normative aim of sustainability to an empirical fact.

The turn to resilience in the water sector has naturalized social and cultural difference to a single, planetary view of the Earth system (Schmidt, 2017). This perspective does not deny social or cultural differences exist, but it does naturalize them to the state of the system in which they operate. Though subtler than its predecessors, resilience is a version of what legal scholar Douglas Kysar (2010) terms "regulating from nowhere" in which the criteria for environmental governance are not referenced to the political community they issue from. For instance, the anthropologist Paul Nadasdy (2007) argues that while resilience-based approaches claim to operate both scientifically and closer to Indigenous notions of conservation, they shift values in ways that reinforce forms of economic accumulation that are premised on the dispossession of Indigenous peoples. A generic point that may be drawn from such critiques is that resilience shifts the register of value from political equality to norms of sufficiency by reckoning resilience in terms of what a social-ecological system needs in order to maintain its functions, identity, and feedbacks. In this sense, resilience is not apolitical. It is both scientific and normative; a way of ordering and valuing social-ecological systems (see MacKinnon and Derickson, 2013). To see how sufficiency-based norms of resilience may also naturalize moral luck, the next section considers how linking the HRWS to resilience has been advanced by calls to revalue water in a new frame of reference: one where water's variability is no longer a stable theatre for human development, but a changing, potentially dangerous force.

Value and the UN High-Level Panel on Water

The UN High-Level Panel on Water attempted a rapprochement between human rights and resilience through the 24-month, "valuing water" initiative launched in 2016. The consultation on valuing water was outlined in a "road map" in May 2017, which argued that values had a special role owing to other UN HLPW (2017a, p. 6) initiatives regarding resilience and climate change, where "the values of water is [sic] (or should be) an important factor for allocation and efficient use of water." A key stop on the global tour of conferences for the valuing water initiative was Bellagio, Italy in May 2017. There, the UN HLPW (2017b, p.1) asserted that water was "precious, fragile, and dangerous." This opening volley was later solidified in the UN HLPW (2017c) outcomes, which moved the focus from water's uneven distribution as the locus for water scarcity to emphasize norms of resilience—fragile tipping points and dangerous extremes—affecting water security. This shift is noteworthy because the UN HLPW's (2018) aim of "making every drop count" explicitly supports the Sustainable Development Goals (SDGs) and, by positioning water's value in relation to resilience and sufficiency, a path was provided to naturalize water needs across sectors, such as food and health, that either directly or indirectly require water.

The HLPW "valuing water" initiative drew on background reports and academic articles to justify its approach to value, and in particular on two background reports from the World Bank. The first was on aid flows to the water sector and a second was on financing options for the 2030 SDG water agenda. These scoped the range and extent of how "water values" organize and mediate competing demands. The discussion of aid flows in the water sector focused attention on the small share of climate finance funds received by the water sector even though water is critical to climate and urban resilience (see Winpenny et al., 2016). The financing options paper highlighted the links between water, efficiency, governance, and creditworthiness in the pursuit of the SDGs (see Kolker et al., 2016). Both reports make routine use of financial products and governance techniques to link water, resilience, and values, such as assessments of creditworthiness, bonds, microfinance, asset management, and insurance. In addition to the World Bank papers, policy forum publications in the journal *Science* framed the challenge of valuing water in terms of the large investments needed to meet the water-related SDGs (see Garrick et al., 2017). Across the various forms of cultural capital that these reports and articles provided to the HLPW's "valuing water" initiative, the integrated nature of economic and environmental concerns was central.

Configured across resilience, finance, and the SDGs, the UN HLPW's (2018, p. 15) concluding documents on water's value use monetary metaphors to argue that, "Water is the common currency which links nearly every SDG, and it will be a critical determinant of success." This mode of connecting water's value to other sectors—as moral and material currency—is critical, according to the UN HLPW (2018, p. 5), because what is at "stake is our human right to access to safe drinking water and sanitation and our future survival." Indeed, the UN HLPW (2018, p. 5) prefaced these stakes with the specter of human impacts on the Earth system, in which climate change ". . . is exacerbating natural variability of the water cycle, increasing water stresses that constrain social progress and economic development." Further, the UN HLPW (2018, p. 22) states that policies should be put in place "consistent with the human right to safe drinking water and the commitment to achieving the SDGs that are affected by water, such as those related to health, poverty, sustainability and resilience." The new alignments of "value" with resilience, finance, and the SDGs is broadly in line with the sufficiency-based norms for the HRWS provided by experts to the UN back in 2010. In this new configuration, however, moral luck is naturalized by shifting the normative aims of human rights to the delivery of sufficient minimums of water—and related development needs—through the particular economic configuration and financial tools of the global economy. That is, non-contingent human rights are made relative to the contingent integration of social-ecological systems, the resilience of which is used to naturalize the pursuit of sufficiency. This form of naturalization does not consider how economic or financial logics may themselves produce inequality, such as through dispossessive forms of property. Instead, moral desert is made relative to the state of an integrated system, the resilience of which must be maintained to avoid collapse across the shared "causal architecture" of economics and environment.

Elsewhere, I have argued that a key normative aspect of resilience-based water management is the idea that environmental and economic systems are *already* integrated (Schmidt, 2017). This proposition shifts the parameters of global sustainability because "integration" no longer functions as the normative aim of sustainable development and is instead taken as a fact describing the mutual coevolution of social and ecological systems. With integration as fact, the normative aim is to facilitate the governance of the many points of connection through which the Earth and the economy are integrated—supply chains of food, water, and energy and the nexus that connects water to other SDGs in frequently unanticipated ways (Schmidt and Matthews, 2018). What this idea fails to take a full moral reckoning of, however, is that the "integration" of environmental and economic systems did not take place on the ground of equality. The problem is compounded when assessments of the needs-based, sufficiency requirements for maintaining the resilience of integrated economic and environmental systems do not address how particular social or cultural forms of integration generate inequality. When extended to the HRWS, the upshot is that the appeal to resilience naturalizes moral luck by shifting values from non-contingent rights of equality to sufficiency-based requirements of the integrated systems individuals and communities happen to live in.

Water and a "right to have rights"

Appeals to resilience, like strategic uses of human rights, can be put to different uses by communities seeking to contest existing distributions of environmental or economic harms and goods. But these strategic uses of resilience depend on first belonging to such a community. In this conclusion, I'm concerned with cases where resilience is used to disarm political communities through the naturalization of moral luck. When moral luck is naturalized it constrains efforts to arrest inequality because the requisite of belonging to a community, the condition Arendt described for a "right to have rights," is made subordinate to considerations of sufficiency. In short, my aim is to add a political qualifier to a standard empirical claim: water is empirically necessary for life, and it is also a prerequisite for belonging to a political community.

Writing a half-century ago, Arendt (1958b) worried that only when individuals have become stateless—bereft of the possibility of belonging to a political community—did a "right to have rights" become apparent. Here, I conclude not with the specter of totalitarianism that occupied Arendt but with a consideration of how belonging to a community is requisite for rights owing to the fact that for a right to be considered legitimate, it must be reckoned as such by a group larger than that of the claimant(s). In short, rights are social. For instance, when the unfolding water crises in Flint, Michigan came before the court, the ruling judge agreed that water was empirically necessary for life but that the city's contaminated water could not be legally resolved in favor of the claimants without an "enforceable right" to water (see DeGooyer et al., 2017). Without recourse to legal rights at the state level of a political community, advocates turned to the norms of a larger political community by appealing to the UN human right to water. Individuals in Flint had not been rendered stateless, but they belonged to a political community that hadn't recognized water is a basic right—water is a prerequisite to a "right to have rights," because, without water, the empirical conditions for political belonging are unavailable.

The tragedy in Flint, not unlike struggles for water elsewhere, unfolded in a context of deep racial and economic inequality (e.g. von Schnitzler, 2016; Chance, 2018; Clark, 2018). Orienting human rights towards equality requires taking up the structural struggle over how belonging is configured within political communities and the distribution of harms and goods across their membership. That is a task that is not well served by the current efforts in global water governance towards norms of "valuing water" that seek resilience-based accounts of sufficiency. The human right to water and sanitation will not be quenched through norms of sufficiency that leave structures of inequality in place because such norms do not solve water challenges but merely displace them. By making the HRWS relative to the resilience of an integrated ecological and environmental system, sufficiency-based norms naturalize moral luck by situating belonging in reference to only one configuration of a political community. Many other forms of life and of political community are possible, however, and for the human right to water and sanitation to fit with the broader aspirations of the Universal Declaration of Human Rights, the pursuit of equality must be central to sharing water among them.

References

Anand, N. (2017) *Hydraulic city: water and the infrastructures of citizenship in Mumbai.* Durham, NC: Duke University Press.
Arendt, H. (1958a) *The human condition.* Chicago, IL: University of Chicago Press.
Arendt, H. (1958b) *The origins of totalitarianism.* New York: Meridian Books.
Arendt, H. (1982) *Lectures on Kant's political philosophy.* Chicago, IL: University of Chicago Press.
Bakker, K. (2012) 'Commons versus commodities: Debating the human right to water (with new postscript),' in Sultana, F. and Loftus, A. (eds), *The right to water: politics, governance, and social struggles.* London: Routledge.
Bhandar, B. (2018) *Colonial lives of property: law, land, and racial regimes of ownership.* Durham, NC: Duke University Press.
Biggs, E., Bruce, E., Boruff, B., Duncan, J., Horsley, J., Pauli, N., McNeill, K., Neef, A., Van Ogtrop, F., Curnow, J., Haworth, B., Duce, S., and Imanari, Y. (2015) "Sustainable development and the water-energy-food nexus: A perspective on livelihoods," *Environmental Science and Policy*, 54, pp. 389–397.
Boelens, R., Perreault, R., and Vos, J. (eds) (2018) *Water justice.* Cambridge: Cambridge University Press.
Boyd, D. (2017) *The rights of nature: A legal revolution that could save the world.* Toronto, ON: ECW Press.

Butler, L. (2000) 'The pathology of property norms: Living with nature's boundaries,' *Southern California Law Review*, 73, pp. 927–1016.

Byrd, J. (2011) *The transit of empire: indigenous critiques of colonialism*. Minneapolis: University of Minnesota Press.

Carpenter, S., Walker, B., AnderiesJ., and Abel, N. (2001) 'From metaphor to measurement: resilience of what to what?,' *Ecosystems*, 4 (8), pp. 765–781.

Chance, K. (2018) *Living politics in South Africa's urban shacklands*. Chicago, IL: University of Chicago Press.

Chatterjee, P. (2006) *The politics of the governed: reflections on popular politics in most of the world*. New York: Columbia University Press.

Clark, A. (2018) *The poisoned city: Flint's water and the American urban tragedy*. New York: Metropolitan Books.

Conca, K. (2015) *An unfinished foundation: the United Nations and global environmental governance*. Oxford: Oxford University Press.

Cooper, M. (2010) 'Turbulent worlds: financial markets and environmental crisis,' *Theory, Culture, and Society*, 27 (2–3), pp. 167–190.

Cooper, M. (2011) 'Complexity theory after the financial crisis,' *Journal of Cultural Economy*, 4 (4), pp. 371–385.

Curle, C. (2007) *Humanité: John Humphrey's alternative account of human rights*. Toronto, ON: University of Toronto Press.

DeGooyer, S., Hunt, A., Maxwell, L., and Moyn, S. (2017) *The right to have rights*. London: Verso.

Eckstein, G. (2010) 'Water scarcity, conflict, and security in a climate change world: Challenges and opportunities for international law and policy,' *Wisconsin International Law Journal*, 27 (3), pp. 410–461.

Falkenmark, M. and Rockström, J. (2004) *Balancing water for humans and nature: the new approach in ecohydrology*. London: Earthscan.

Feldman, D. (2007) *Water policy for sustainable development*. Baltimore, MD: John Hopkins University Press.

Folke, C. (2003) 'Freshwater for resilience: a shift in thinking,' *Philosophical Transactions of the Royal Society of London B*, 358, pp. 2027–2036.

Folke, C. (2006) 'Resilience: the emergence of a perspective for social-ecological systems analyses,' *Global Environmental Change*, 16, pp. 253–267.

Galaz, V. (2007) 'Water governance, resilience and global environmental change: a reassessment of integrated water resources management (IWRM),' *Water Science and Technology*, 56 (4), pp. 1–9.

Garrick, D., Hall, J., Dobson, A., Damania, R., Grafton, R.Q., Hope, R., Hepburn, C., Bark, R., Boltz, F., de Stefano, L., O'Donnell, E., Matthews, N., and Money, A. (2017) "Valuing water for sustainable development," *Science*, 358 (6366), pp. 1003–1005.

Global Water Partnership (2012) *Water security and climate resilient development: technical background document*. Stockholm: Global Water Partnership.

Hall, A. (2010) *Earth into property: colonization, decolonization, and capitalism*. Montréal, QC: McGill-Queens University Press.

Hellum, A., Mbote-Kameri, P., and van Koppen, B. (eds) (2015) *Water is life: women's human rights in national and local water governance in Southern and Eastern Africa*. Harare: Weaver Press.

Homer-Dixon, T., Thomas, B.W., Biggs, R., Crepin, A.-S., Folke, C., Lambin, E., Peterson, G.D., Rockström, J., Scheffer, M., Steffen, W., and Troell, M. (2015)

"Synchronous failure: The emerging causal architecture of global crisis," *Ecology and Society*, 20 (3), art 6.

Jamieson, D. (2014) *Reason in a dark time: why the struggle against climate change failed—and what it means for our future*. New York: Oxford University Press.

Kolker, J., Kingdom, B., Trémolet, S., Winpenny, J., and Cardone, R. (2016) *Financing options for the 2030 water agenda*. Washington, DC: World Bank.

Kysar, D. (2010) *Regulating from nowhere: environmental law and the search for objectivity*. New Haven, CT: Yale University Press.

Langford, M. and Russell, A. (eds) (2017) *The human right to water: theory, practice, and prospects*. Cambridge, MA: Cambridge University Press.

Macekura, S. (2015) *Of limits and growth: the rise of global sustainable development in the twentieth century*. Cambridge, MA: Cambridge University Press.

MacKinnon, D. and Derickson, K. (2013) 'From resilience to resourcefulness: a critique of resilience policy and activism,' *Progress in Human Geography*, 37 (2), pp. 253–270.

Milly, P.C.D., Betancourt, J., Falkenmark, M., Hirsch, R.M., Kundzewicz, Z.W., Lettenmaier, D.P., and Stouffer, R.J. (2008) "Stationarity is dead: whither water management?," *Science*, 319, pp. 573–574.

Moyn, S. (2018) *Not enough: human rights in an unequal world*. Cambridge, MA: Harvard University Press.

Nadasdy, P. (2007) 'Adaptive co-management and the gospel of resilience,' in Armitage, D., Berkes, F., and Doubleday, N. (eds) *Adaptive co-management: collaboration, learning, and multi-level governance*. Vancouver, BC: UBC Press.

Pahl-Wostl, C. (2015) *Water governance in the face of global change: from understanding to transformation*. Dordrecht: Springer.

Perera, V. (2015) 'Engaged universals and community economies: the (human) right to water in Colombia,' *Antipode*, 47 (1), pp. 197–215.

Rockström, J., Falkenmark, M., Allan, J.A., Folke, C., Gordon, L., Jägerskog, A., Kummu, M., Lannerstad, M., Meybeck, M., Molden, D., Postel, S., Savenije, H.H.G., Svedin, U., Turton, A., and Varis, O. (2014) "The unfolding water drama in the Anthropocene: towards a resilience-based perspective on water for global sustainability," *Ecohydrology*, 7, pp. 1249–1261.

Sachs, J. (2015) *The age of sustainable development*. New York: Columbia University Press.

Schmidt, J. (2012) 'Scarce or insecure? The right to water and the ethics of global water governance,' in Sultana, F. and Loftus, A. (eds) *The right to water: politics, governance, and social struggles*. London: Routledge.

Schmidt, J. (2017) *Water: abundance, scarcity, and security in the age of humanity*. New York: New York University Press.

Schmidt, J. and Matthews, N. (2018) 'From state to system: financialization and the water-energy-food-climate nexus,' *Geoforum*, 91, pp. 151–159.

Schmidt, J. and Mitchell, K. (2014) 'Property and the right to water: toward a non-liberal commons,' *Review of Radical Political Economics*, 46 (1), pp. 54–69.

Shue, H. (1980) *Basic rights: subsistence, affluence, and US foreign policy*. Princeton, NJ: Princeton University Press.

Shue, H. (2014) *Climate justice: vulnerability and protection*. Oxford: Oxford University Press.

Steffen, W., Sanderson, A., Tyson, P.D., Jäger, J., Matson, P., Moore, B., Oldfield, F., Richardson, K., Schellnhuber, J., Turner, B., and Wasson, R. (2004) *Global change and the Earth system: a planet under pressure*. Berlin, Germany: Springer.

Steffen, W., Rockström, J., Richardson, K., Lenton, T., Folke, C., Liverman, D., Summerhayes, C., Barnosky, A., Cornell, S., Crucifix, M., Donges, J., Fetzer, I., Lade, S., Scheffer, M., Winkelmann, R., and Schellnhuber, H.J. (2018) "Trajectories of the Earth system in the Anthropocene," *Proceedings of the National Academy of Sciences*, 115, pp. 8252–8259.

United Nations General Assembly (1948) *Resolution 217 A: Universal Declaration of Human Rights*. Paris: United Nations. Available at: www.un-documents.net/a3r217a.htm

United Nations General Assembly (2010) *Resolution A/65/254: Human rights obligations related to access to safe drinking water and sanitation*. New York: United Nations.

United Nations High-Level Panel on Water (2016) *Background note*. pp. 1–3. Available at: sustainabledevelopment.un.org/HLPWater

United Nations High-Level Panel on Water (2017a) *Roadmap of the valuing water initiative*. United Nations High-Level Panel on Water. Available at: https://sustainabledevelopment.un.org/content/documents/15595Road_Map_Valuing_Water_Initiative_vs_1.1_March_8th_updated_May_19th_2017.pdf

United Nations High-Level Panel on Water (2017b) "Bellagio principles on valuing water, *United Nations* " pp. 1–3. Available at: https://sustainabledevelopment.un.org/content/documents/15591Bellagio_principles_on_valuing_water_final_version_in_word.pdf

United Nations High-Level Panel on Water (2017c) *Value water*. United Nations High-Level Panel on Water. Available at: https://sustainabledevelopment.un.org/content/documents/hlpwater/07-ValueWater.pdf

United Nations High-Level Panel on Water (2018) *Making every drop count: an agenda for water action*. Available at: https://sustainabledevelopment.un.org/HLPWater

von Schnitzler, A. (2016) *Democracy's infrastructure: techno-politics and protest after apartheid*. Princeton, NJ: Princeton University Press.

Walker, J. and Cooper, M. (2011) 'Genealogies of resilience: From systems ecology to the political economy of crisis adaptation,' *Security Dialogue*, 42 (2), pp. 143–160.

Winpenny, J., Trémolet, S., and Cardone, R. (2016) *Aid flows to the water sector: overview and recommendations*. Washington, DC: World Bank.

3 Making space for practical authority
Policy formalization and the right to water in Mexico

Katie Meehan

A new challenge

Safe and secure access to water and sanitation is widely recognized as a fundamental human right, enshrined in the Dublin Principles and promoted as a key point of policy reform by international organizations, civil society groups, and government actors. In 2012, Mexico amended its national constitution to recognize the right of people in Mexico—regardless of national origin—to safe, sufficient, affordable, and reliable water and sanitation for personal and domestic use. On the heels of similar reforms in Bolivia, Chile, and Uruguay, Mexico's constitutional reform was widely celebrated in the Americas as a milestone in promoting a human-rights-based approach to water governance.

The reform touched down in a country marred by persistent disparities in household water provision and security. In 2015, an estimated 55 percent of Mexican households counted piped water inside their dwellings, 36 percent of households had water connections within their building or plot of land, and 9 percent of households lacked piped water entirely (Instituto Nacional de Estadística, Geografía e Informática, 2015). Rural, indigenous, and peri-urban communities are particularly vulnerable to problems of water insecurity: such as poor quality, unreliable service, and exploitation at the hands of private water vendors and municipal providers who exchange piped water for votes (Castro, 2006; Castro, Heller, and da Piedade Morais, 2015; Herrera, 2017; Pacheco-Vega, this volume).

Even households with piped connections encounter water shortages and problems. In Mexico City, the second-largest metropolis in the hemisphere, roughly 20 percent of city residents cannot count on daily tap water supply. Plumbed households must wait their turn for tap water (a rationing system called *el tandeo*) or purchase water deliveries from tanker trucks (*pipas*). In some areas, pack animals literally carry the burden of a dysfunctional public network. "A marginalized neighborhood in Mexico City depends on the use of donkeys to transport water," reports Léo Heller, the Special Rapporteur of the United Nations Human Rights Council (Office of the United Nations High Commissioner for Human Rights [UN OHCHR], 2017), "while other communities of the City reported that the water in their localities is diverted to high commercial, residential and tourist uses."

Despite the bright promise of constitutional reform, the Special Rapporteur's report signals a new challenge for the human right to water. In recent years, the policy challenge has moved beyond formal acknowledgement of the right to water and sanitation—in constitutional language, judicial decisions, or UN proclamations—to the uneven geography of policy implementation, enforcement, and practice on the ground (Baer and Gerlak, 2015; Olmos Giupponi and Paz, 2015; Baer, 2017). Located in the grey zone between rights talk and its realization is an urgent question regarding the authority of new institutions: how do policy ideas—like the human right to water—gain traction in real life?

In this chapter, I draw on the case of Mexico to explore how *practical authority* for the human right to water is developed through a diverse and sometimes unexpected set of institutional actors, sites, and strategies. Practical authority is a "kind of power-in-practice generated when particular actors (individuals or organizations) develop capabilities and win recognition within a particular policy area, enabling them to influence the behavior of other actors" (Abers and Keck, 2013, p. 2). Policy formalization involves the two-way transformation of social values and norms into formal rules and rights, a process that adds flesh and blood to the bones of policy narratives and ideas (Meehan and Moore, 2014). Yet to formalize policy, new institutions like the human right to water require practical authority: "a kind of power in which the capabilities to solve problems and [win] recognition by others allows an actor to make decisions that others follow" (Abers and Keck, 2013, p. 7).

Rather than focus on the usual suspects of policy formalization—legal texts, elected officials, and government agencies—in this chapter, I examine how practical authority is developed at the margins of the state: the experimental spaces where governance ideas gain skills, capacities, and legitimacy on the ground (Abers and Keck, 2013; Angel and Loftus, 2019; McConnell, 2017). Drawing on a practice-based approach to institutional change, I identify three pathways that actors use to navigate policy impasse and to build practical authority for the human right to water. A perspective from the margins permits a more robust and comprehensive understanding of how novel policy ideas—such as the human right to water—come to life in complex institutional settings, in and beyond the official state apparatus (McConnell, Moreau, and Dittmer, 2012; Jeffrey, McConnell, and Wilson, 2015; Angel and Loftus, 2019; McConnell, 2017).

Practical authority is especially important in a situation of policy impasse. In Mexico, progress in developing federal regulatory instruments and enforcement mechanisms for the right to water has stalled since 2012. Despite the impasse, I show how nonstate actors have maintained important pressure and visibility for the right to water, innovated new service delivery models, and introduced alternative pathways of policy implementation at different scales. I argue that a practice-based approach reveals the diverse pathways, spaces, and capabilities needed to make institutional change permanent—to realize rights talk in daily life, beyond its constitutional script.

Mexico is an important proving ground for realizing human rights policy. Over the past decade, the country has experienced a stark increase in kidnapping and torture cases, organized crime and gun violence, police brutality, assault and detention of migrants, gender-based violence, and the murder of journalists. In 2013, a Human Rights Watch report declared the country to harbor the most severe crisis of enforced disappearances in Latin America in decades (in Wright, 2018). Entrenched problems of elite impunity, corruption, clientelism, and neoliberal policy reforms have sparked fresh outrage and public protest (Wright, 2017, 2018). Domestic water insecurity is a less immediate threat to human rights, but its presence as a form of institutionalized violence is no less pervasive. In a context of extreme public mistrust in authorities, this chapter suggests that actors increasingly rely on diverse tactics and strategies to build practical authority for human rights, especially at the margins of state power.

My analysis of the formalization process draws on a variety of data sources: interviews with institutional actors, reports and press releases, unpublished materials by NGOs, participant observation of water activists, and legal analysis and documents produced by Centro Mexicano de Derecho Ambiental (CEMDA). In contrast to conventional policy analysis, which seeks to identify ingredients that lead to "success" or "failure" of a policy idea, my approach advances a practice-based understanding of institutional transformation and change—one that situates policy as a process of creative world-making, in and beyond the formal halls of power.

Institutionalizing the human right to water in Latin America

Policy ideas and narratives travel through space; they also secrete and come to life in the thick of particular places. For nearly half a century, Latin America has been a testing ground for new policy ideas and development interventions in the water sector (Meehan, 2019). Starting in the 1970s under the Pinochet regime, Chile introduced the world to market-based water governance by reforming laws and regulations to inject free-market principles in water rights adjudication (Budds, 2004). The so-called "Chilean miracle" spread to other countries in the hemisphere, such as Peru, which adopted similar market-friendly water policies and investment strategies during the Fujimori government (Seemann, 2016).

In Latin America, the human right to water emerged as a policy narrative in response to neoliberal reforms and water privatization in the late 1990s and 2000s. To date, Bolivia, Ecuador, Mexico, Peru, and Uruguay have adopted the right to water as constitutional principles and national policy mandates (Bustamante, Crespo, and Walnycki, 2012; Harris and Roa-García, 2013; Seemann, 2016; Baer, 2017; Roa-Garcíaa, Urteaga-Crovetto, and Bustamante-Zenteno, 2017). The constitutions of Chile, Guatemala, and Venezuela include an indirect recognition of the right to water. Beyond constitutional reforms, courts in

Argentina and Colombia have upheld the right to water through judicial decisions (Olmos Giupponi and Paz, 2015).

While the constitutional right to water is upheld by some scholars as evidence of resistance to neoliberal principles of water governance (Harris and Roa-García, 2013), this point is under some debate (Bustamante, Crespo, and Walnycki, 2012; Seemann, 2016; Baer, 2017). For example, Madeline Baer and Andrea Gerlak (2015, p. 1531) argue that "the human rights frame does not necessarily challenge the dominant knowledge about the merits of marketization of water that is at the core of neoliberal water policy." In the Americas, scholars tend to understand the human right to water in terms of formal institutional arrangements. As Baer (2017) argues, a legal-centric focus risks obscuring the right to water as a dynamic and fundamentally *social* process, subject to a variety of institutional actors and organizations operating at different scales, abilities, and capacities.

A practice-based approach opens up different modalities of explanation. In their decade-long study of Brazilian water policy reform, Rebecca Neaera Abers and Margaret Keck (2013) advance a practice-based approach to institutional change and policy formalization. A key component of their approach is institutional entanglement: the idea that "overlapping administrative jurisdictions layered upon ambiguous functional divisions of labor may produce competition for, confusion about, or even gaps in political authority" (Abers and Keck, 2013, p. 21). Federalist countries, such as Brazil and Mexico, are especially prone to institutional entanglement. As Abers and Keck (2013) explain, legal and regulatory overlap are also important points of intervention, particularly for non-state actors. In Brazil, institutional entanglement in the water sector led to a proliferation of new civic groups and organizations, exchanges and networks, and opportunities for state-society interaction (Abers and Keck, 2013, p. 39).

Practical authority is achieved through a variety of tactics and strategies. In the sections to follow, I elaborate three techniques used by actors in Mexico: expert pressure, hedging, and practical experimentation. Together, these practices keep the right to water alive on the policy agenda and in the public domain—a beacon in the context of policy stalemate, human rights violations, and extreme water insecurity.

Making space for practical authority in Mexico

In 2012, the Mexican Congress approved reforms to the national constitution to recognize the human right to water and sanitation (Table 3.1). The Congress amended the language of several constitutional articles. Article 1 establishes the legal regime for the implementation of human rights. Article 4 establishes the human right to water and sanitation, as well as the right of people to live in a healthy environment for their development and well-being. Article 4 calls on federal, state, and municipal authorities to recognize the right to water, and holds accountable actors who cause damage or environmental destruction (CEDMA-ELI, 2016). In addition to these reforms, two additional water-related provisions in the constitution are Article 27 (establishes state territory and the legal regime of national waters) and Article 115 (delineates the powers of municipalities to provide local public services such as drinking water, sewerage, and wastewater treatment) (CEDMA-ELI, 2016).

Adapted from CEDMA-ELI (2016) and Rabasa et al. (2014), with updated information from the UN Human Rights Council (2017)

Constitutional reforms stipulate that domestic consumption of water and sanitation should be provided in a way that is safe, sufficient, acceptable, and affordable (Rabasa, Velasco, and Martínez, 2014). The Judicial Branch has determined that Article 4 is a subjective right, which means it may be enjoyed by any person within Mexico, regardless of their national origin or background (Rabasa, Velasco, and Martínez, 2014). One of the most important implications of the Article 1 reform, as interpreted by the Supreme Court of Mexico, is expansive protection of human rights, including international treaties signed by the federal government (Rabasa, Velasco, and Martínez, 2014).

Despite the constitutional success, implementation has stalled. To date, Mexico lacks any regulatory law, enforcement mechanisms, or independent oversight to ensure the right to water, resulting in a constitutional mandate without the necessary policy teeth for effective, equitable, and comprehensive realization (Rabasa, Velasco, and Martínez, 2014; UN OHCHR, 2017). For example, neither the National Waters Law (first established in 1992, updated in 2008) nor state-level drinking water laws and municipal provision ordinances have been updated to reflect the human right to water. The National Waters Law is the most influential and important institutional framework for managing water in Mexico. Yet, in its current iteration, the Law facilitates private sector involvement and a public consultation process that has done little to ameliorate service problems and equity gaps in water provision (Wilder, 2008, 2010).

In the face of this policy impasse, institutional actors in Mexico have struggled to carve different pathways toward policy formalization. As the following anecdotes demonstrate, making space for practical authority requires creative action and careful navigation around existing institutional arrangements and policy impediments.

Expert pressure

Pressure makes water flow. Through forces both physical and social, actors use pressure tactics to make water available to diverse groups and geographies. "Pressure can be mobilized by using pumps or politicians," writes Nikhil Anand (2011, p. 543), "and access to the technologies of pressure is mediated as much by capital as by social connections." Pressure works in multiple directions. For example, local actors in Mumbai use a variety of tactics to compel decision makers and utility operators into better water service (Anand, 2011). In this section, I use the term *expert pressure* to describe how elite global actors follow prescribed lines of influence to develop practical authority for institutional change.

Table 3.1 Policy instruments applicable to the human right to water in Mexico

Instrument	Function	Source	Notes
Article 1	Establishes the general regime of protection of human rights	National constitution	Authorities have an obligation to implement human rights by observing principles of universality, interdependence, indivisibility, and progressivity.
Article 4	Establishes the human right to water and sanitation; establishes the right of all people to live in a healthy environment; federal, state, and municipal authorities should recognize this right; recognizes responsibility of those who cause damage or environmental destruction	National constitution	Decisions related to water provision and infrastructure development should consider and protect ecosystem services; the Judiciary specifies Article 4 as a "subjective" or universal right and, therefore, applicable to any person within national territory.
Article 25	The state must guarantee that development is integral and sustainable	National constitution	Decisions should also consider social and economic factors, the environment, and intergenerational needs.
Article 27	Establishes legal regime and national dominion over territorial waters.	National constitution	Changes to Article 4 did not modify the legal framework of national waters or territory, as explained in Article 27, paragraph five
Article 115	Municipalities are tasked with the responsibility to provide public services for drinking water, sewerage, and treatment/disposal of wastewater.	National constitution	Changes to Article 4 did not modify the legal framework of municipal decentralization or responsibility for public service provision.
National Water Law	Stipulates federal regulations	Regulatory law	Passed in 1992; revized in 2008. This law is the single most important legal instrument for water in Mexico. As of 2018, the law does not incorporate a human rights framework; nor does it reflect 2012 amendments to the Constitution.
Water Sustainability Law	Adopts a human rights approach to water management and regulation in Mexico City	Regulatory law	Passed in 2014; contains controversial concessions to private sector involvement; transforms public utility into an autonomous unit; applies to Mexico City only

Data source: adapted from CEDMA-ELI (2016) and Rabasa et al. (2014), with updated information from the UN Human Rights Council (2017).

A prime example of expert pressure is the Special Rapporteur for the Human Right to Water and Sanitation. Established in 2008 by the UN Human Rights Council (Resolution 7/22), the Special Rapporteur is charged with providing policy guidance, monitoring progress and impediments, and identifying best practices to implement safe drinking water and sanitation in countries around the world. Within the UN system, Special Rapporteurs are independent experts appointed by the UN Human Rights Council. Catarina de Albuquerque served as the first Special Rapporteur for the Human Right to Water and Sanitation from 2008 to 2014; Léo Heller was appointed in 2014. With a goal of shaping policy discourse, the Special Rapporteur makes official visits to assess the situation of water and sanitation in various countries (Baer and Gerlak, 2015). Between 2009 and 2017, the Special Rapporteur participated in 21 country visits.

In May of 2017, the Special Rapporteur made an official visit to Mexico. Following a Papal visit to Mexico in February 2016—Pope Francis is a vocal supporter of the human right to water—the Special Rapporteur's trip to Mexico marked the second visit in under two years by a prominent international human rights advocate. Over 11 days, Mr. Heller met with representatives of the federal government, state and municipal authorities, and civil society organizations. He made visits to urban and rural communities in central and southern Mexico, speaking directly with residents about their provision challenges.

Using his international platform, the Special Rapporteur identified legal gaps and institutional weaknesses in the Mexican right to water. In addition to the outdated National Water Law, he noted a lack of financial resources and technical capacity in municipal governments, which are ultimately responsible for providing water and sanitation services. He highlighted the absence of an independent regulatory body, as private and social actors in water provision—such as informal water sellers, community water boards, bottled water vendors, and sanitation providers—are currently unregulated in Mexico (Rabasa, Velasco, and Martínez, 2014; UN OHCHR, 2017).

Beyond legal issues, the Special Rapporteur made a direct and powerful analogy between water insecurity and other forms of systemic violence. "Civil and political rights issues including addressing allegations of torture and forced disappearances were noted as high priorities of concern to Mexico," he wrote (UN OHCHR, 2017), "While this is vital, I encourage the Government, as required under international human rights law and standards, to give equal and appropriate attention and to provide necessary resources to addressing critical economic, social and cultural rights, including the rights to safe drinking water and sanitation."

As a tactic that follows prescribed lines of influence, expert pressure has recognizable limitations. "[The Special Rapporteur] approach may be effective in getting states to recognize their crucial role in the water sector," argue Baer and Gerlak (2015, p. 1538), "but it is not a transformative discourse that challenges old paradigms or power structures." For example, previous reports by the Special Rapporteur have been reluctant to oppose private sector participation in water services (Baer and Gerlak, 2015). The Mexico visit continues this pattern. Bottled water consumption and informal vendors are noted as areas of concern, but the Special Rapporteur's report is mute on the question of whether private sector involvement could even guarantee safe, sufficient, acceptable, and affordable water and sanitation (Pacheco-Vega, this volume). In this way, expert pressure risks promoting a vehicle of change without a paradigm shift.

Nonetheless, the Special Rapporteur lends important elements of international visibility, credibility, and political salience to the human right to water in Mexico. Expert pressure may leverage the positional power of policy actors, who are lodged in bureaucratic hierarchies (Abers and Keck, 2013). In contrast to Pope Francis, who casts the right to water in moral tones, the Special Rapporteur helps to shape the discourse of policy implementation as an issue of *governance* in Mexico, not ethical will.

Hedging

Compared to expert pressure, which follows prescribed lines of influence, *hedging* describes a process in which actors construct policy implementation possibilities in more than one institutional arena at the same time. "They are, proverbially, putting their eggs in more than one basket," explain Abers and Keck (2013, p. 25), "under conditions of substantial uncertainty about the future of any given effort." Hedging is a response to overlapping jurisdictions, institutional entanglement, and uncertainty about future outcomes. "In the Piracicaba basin," explain Abers and Keck (2013, p. 202), "a group of water specialists allied with businesses and mayors managed to dominate three overlapping arenas at the same time: the consortium they had created, a river basin committee created under state law, and another created under federal law." Unsure about policy outcomes but wanting to stay in the game, hedgers help to build practical authority in multiple and even competing institutional domains.

In Mexico, an example of hedging is the Water Sustainability Law (Ley de Sustentabilidad Hídrica) of Mexico City. First introduced in 2014, the law was passed by the city's Legislative Assembly in November 2017 with 33 votes in favor, 14 votes in opposition, and one abstention. The law requires the city government to conform to the principles of human rights, sustainability, transparency, and shared responsibility. Notably, the law provides legal, regulatory, and enforcement measures for the human right to water and sanitation—not currently provided by the National Water Law—for residents within the geographic limits of Mexico City. Implementation of the Water Sustainability Law must conform to the principles of universality, interdependence, indivisibility, and progressivity—the same guidelines stipulated by Article 4 of the Federal Constitution.

The new Law introduces several controversial elements that have generated opposition from human rights organizations and local activists. For example, the law has transformed the legal status of the local water utility, known as

Sacmex (Sistema de Aguas de la Ciudad de México). Under the new law, Sacmex has full technical, budgetary, and management autonomy, including the ability to sign third-party contracts with private companies for the provision of water and sanitation services—a sticking point that caused legislators from opposition parties to debate and delay the vote on the bill (including Andrés Manuel López Obrador, a populist politician, former Mexico City mayor, and current President of Mexico). Among its other controversial elements, the bill also (1) introduces the possibility of a market for harvested and purified rainwater, (2) prohibits the provision of piped water services to irregular settlements on conservation land, and (3) permits the suspension of water service to those who do not pay in two or more consecutive periods (although a "humanitarian supply" of 50L per person per day is guaranteed).

Hedging in Mexico reveals the contested nature of policy narratives, namely the role of private sector participation in the right to water. By hedging in multiple institutional arenas, policy actors strive to build support and consensus for their normative ideal of human rights—such as the idea that private business has no business in ensuring the right to water. For example, a national coalition of water activists, called Coordinadora Nacional Agua Para Tod@s, came together to develop the Citizen's National Water Law: a set of governance principles and decision-making structures to realize the human right to water and sanitation in Mexico. The group presented the first version of the law to federal legislators in 2015 and has since traveled around the country to national forums, university conferences, and public events.

A cornerstone of their approach is the idea that water and its management should remain in the public domain. In contrast to the aforementioned water laws in Mexico, the Citizen's Law explicitly prohibits any institutional arrangement that makes water a commodity, bans private sector control over water resources, and forbids the extraction of profit from any aspect of water management. Elements of this law reflect broader debates about the right to water in Latin America, specifically concerning the encroaching privatization/commercialization of public water services (Castro, Heller, and da Piedade Morais, 2015). While national constitutions have been an important venue where this debate plays out (Harris and Roa-García, 2013), a practice-based approach reveals how authority for the right to water is developed and contested in the full spectrum of institutional arenas.

Hedging is often seen as a response to institutional entanglement. In Mexico, hedging strategies are a clear response to policy debates and deadlock at the federal level. Water in Mexico has a history of strong federal control and clear lines of centralized authority, embodied by the National Water Commission (CONAGUA) and its powerful regulatory scope and mission (Wilder, 2008, 2010). Until the federal government takes more concrete policy actions, institutional actors in Mexico will continue to build practical authority and governance capacity at different scales of government—such in state or city governments—and in multiple institutional arenas, including at the nexus of state/civil society.

Practical experimentation

Creative action is a hallmark of practical authority. To find pathways through institutional entanglement (or impasse), actors may first experiment at a small or local scale, where competition for authority is not strong. *Practical experimentation* involves a combination of ideas, resources, technologies, and relationships in new ways to solve seemingly intractable problems (Abers and Keck, 2013, p. 17). By starting small and resolving localized problems, actors produce intermediate outcomes that test ideas, build trust, develop the organizational capacity for future collaboration, and win external recognition. Federal institutions are slow to change; powerful institutional actors are reluctant to cede their authority. "Despite such asymmetries—or perhaps because of them—experimenting at the small scale involves a great deal of perseverance by actors who [are] otherwise disadvantaged politically" (Abers and Keck, 2013, p. 24).

In Mexico City, the civil society organization Isla Urbana is a novel example of practical experimentation for the sociotechnical realization of the right to water. Formed in 2009, Isla Urbana aims to "catalyze a rainwater harvesting revolution" through the design and implementation of urban rainwater harvesting systems for household use in marginalized communities. Their mission statement is to enhance sustainable water alternatives while reducing water insecurity in the domestic sphere. Their operations and work are financially supported by contracts with city governments and public agencies, by small grants from private and public foundations, and by the sale of their harvesting technologies to more affluent customers. Over the past decade, Isla Urbana has earned awards and recognition from local and international organizations, such as the Government of Mexico City, the Clinton Global Initiative, and the United Nations.

Household water users in Mexico City, particularly in the southern and eastern districts, suffer from sporadic delivery, poor quality service, or the complete absence of a piped connection. Starting in the borough of Tlalpan, Isla Urbana designed a rooftop rainwater collection system that integrates into existing housing stock, seasonal rainfall patterns, water use customs, and household labor practices. In a typical rainy season (July–September) in Mexico City, the rainwater systems supply up to 65 percent of total household water demand, replacing water otherwise purchased from private sources and informal vendors (an unregulated and exploitative market). Rather than replace public water, Isla Urbana aims to transform the sociotechnical mechanism of public service delivery through an idiom of *infrastructural coexistence*: piped water and alternative water provisioning systems existing side by side (Furlong, 2014).

Initially, Isla Urbana started at the periphery of state power: as a young, three-person team equipped with personal drive, plumbing experience, and a background in industrial design. Through their organizational evolution, Isla Urbana exemplifies the trajectory of experimental spaces and technologies developed by actors who desire to advance more sustainable alternatives to the existing sociotechnical regime (Lawhon and Murphy, 2012). By starting small,

Isla Urbana was able to test, modify, and improve their designs and implementation strategies in a relatively short time frame. Since 2009, Isla Urbana has installed over 7,000 domestic harvesting systems and trained 49,938 water users, mostly in partnership with the Mexico City borough governments of Tlalpan, Xochimilco, and Iztapalapa. Their relatively modest size and position as a civil society organization permit speed and institutional flexibility in the water sector. Following the 2016 earthquake, Isla Urbana rapidly installed systems in 101 households that experienced neighborhood water main breaks.

Experiments such as Isla Urbana test the human right to water in important ways. While governments largely agree that water and sanitation should be safe, sufficient, affordable, and reliable, the *technological* pathways of implementation and realization are rarely discussed in human rights circles. Indeed, the "modern infrastructural ideal" of universal, integrated, single-system infrastructure provision is still the design norm in water service delivery models, even in the global South (Furlong, 2014). Thinking beyond the networked ideal requires a context of experimentation that is often best suited to nonstate actors, like Isla Urbana. As with other examples of municipal-supported rainwater harvesting projects (Furlong, 2014), the point of the Isla Urbana system is *not* to replace the grid with individualized collection systems but to modify household elements of public infrastructure in ways that improve user experience and household security. In this way, practical experimentation has the potential to spark a paradigm shift in service delivery models that guarantee and actually provide the right to water.

Undoubtedly, there are also great challenges to practical experimentation in Mexico. Networked infrastructures, such as water and sewerage, are prone to monopolistic control and problems of clientelism. In other words, local politicians "sell" piped water access for votes and party loyalty (Herrera, 2017), including the rainwater systems. In one interview, an Isla Urbana employee describes his awe at patronage politics and how authorities use water infrastructure provision to exert control:

> The interesting thing is the socio-political side. Of course, who gets the systems should be based on necessity, to help people with the highest need. But the politics are in there. They [elected municipal leaders] are giving the [rainwater] systems to the people who voted for them, basically. What I love about it is that they are super organized [laughs]. When we go into [a community], they have lists and we get divided up into teams with a leader who takes you around to the houses. It's like a top-down organization. It's a way that they [municipalities] *massively* send social programs out.

As this comment reveals, experiments like Isla Urbana may solve one aspect of an intractable problem (old service delivery model), only to confront another (clientelism).

The "outsider" role of a nonstate actor has plenty of drawbacks in building practical authority. Isla Urbana struggles with a lack of funding for social innovation and systems designed for secure livelihoods, not profit. As the Water Sustainability Law makes apparent, social technologies can be co-opted by private companies and transformed into revenue-generating streams. Ultimately, the right to water is a constitutional mandate and public policy obligation in Mexico. Therefore, any intermediate outcomes generated by Isla Urbana—such as an alternative service delivery model based on securing water needs—must also navigate and withstand institutionalized hierarchies and existing arrangements. Practical experiments must endure time and competing power interests in order to accrue the capabilities and recognition necessary to transform ideas into policy practice.

Conclusion

In recent years, the uneven geography of policy formalization has emerged as a central challenge for the human right to water. In this chapter, I employed a practice-based approach to explore how the right to water has gained practical authority at different scales and sites in Mexico, a country where federal progress in institutionalizing the right to water has stalled since 2012. This analysis imparts key lessons for the human right to water. First, this chapter provides analytical tools to make sense of policy impasse and political strategies in human rights implementation. A practice-based approach reveals the wide range of actors, organizations, and sociotechnical capabilities needed to make institutional change permanent—to realize rights talk in daily life, beyond its constitutional script.

Second, this chapter demonstrates how actions at the margins of state power are vital in building practical authority for policy implementation. Practical authority is gained through multiple strategies, scales, and spheres of influence. What the case of Mexico reveals is the sheer diversity of efforts to democratize water management and work toward water justice—including top-down and bottom-up approaches, from within and outside the state apparatus, through elite tactics and unorthodox strategies. Constitutional reforms are undoubtedly necessary and important to lasting institutional transformation, but they are not the only sites where progress toward a right to water is exercised, influential, or made real.

References

Abers, R.A. and Keck, M.E. (2013) *Practical authority: agency and institutional change in Brazilian water politics*. New York: Oxford University Press.

Anand, N. (2011) 'PRESSURE: the politechnics of water supply in Mumbai,' *Cultural Anthroplogy*, 26 (4), pp. 542–564.

Angel, J. and Loftus, A. (2019) 'With-against-and-beyond the human right to water,' *Geoforum*, 98 (1): 206–213. doi:10.1016/j.geoforum.2017.05.002

Baer, M. (2017) *Stemming the tide: human rights and water policy in a neoliberal world*. New York: Oxford University Press.

Baer, M. and Gerlak, A. (2015) 'Implementing the human right to water and sanitation: a study of global and local discourses,' *Third World Quarterly*, 36 (8), pp. 1527–1545.

Budds, J. (2004) 'Power, nature, and neoliberalism: the political ecology of water in Chile,' *Singapore Journal of Tropical Geography*, 25 (3), pp. 322–342.

Bustamante, R., Crespo, C., and Walnycki, A.M. (2012) 'Seeing through the concept of water as a human right in Bolivia,' in Sultana, F. and Loftus, A. (eds) *The right to water: politics, governance, and social struggles*. London and New York: Earthscan, pp. 223–240.

Castro, J.E. (2006) *Water, power, and citizenship: social struggle in the basin of Mexico*. London: Palgrave Macmillan.

Castro, J.E., Heller, L., and da Piedade Morais, M. (2015) *O Direito à Água como Política Pública na América Latina*. Brasília: Instituto de Pesquisa Econômica Aplicada.

Centro Mexicano de Derecho Ambiental (CEDMA) and Environmental Law Institute. (2016) *Modelo de Plan de Implementación Municipal del Derecho Humano al Agua*. Mexico City: CEDMA.

Furlong, K. (2014) 'STS beyond the "modern infrastructure ideal": extending theory by engaging with infrastructure challenges in the South,' *Technology in Society*, 38, pp. 139–147.

Harris, L.M. and Roa-García, M.C. (2013) 'Recent waves of water governance: Constitutional reform and resistance to neoliberalization in Latin America (1990–2012),' *Geoforum*, 50 (1), pp. 20–30.

Herrera, V. (2017) *Water and politics: clientelism and reform in urban Mexico*. Ann Arbor: University of Michigan Press.

Instituto Nacional de Estadística, Geografía e Informática (INEGI) (2015) *Intercensal Survey 2015*. Mexico City, Mexico: INEGI.

Jeffrey, A., McConnell, F. and Wilson, A. (2015) 'Understanding legitimacy: perspectives from anomalous geopolitical space,' *Geoforum*, 66 (1), pp. 177–183.

Lawhon, M. and Murphy, J.T. (2012) 'Socio-technical regimes and sustainability transitions: insights from political ecology,' *Progress in Human Geography*, 36 (3), pp. 354–378.

McConnell, F. (2017) 'Liminal geopolitics: the subjectivity and spatiality of diplomacy at the margins,' *Transactions of the Institute of British Geographers*, 42 (1), pp. 139–152.

McConnell, F., Moreau, T. and Dittmer, J. (2012) 'Mimicking state diplomacy: the legitimizing strategies of unofficial diplomacies,' *Geoforum*, 43 (1), pp. 804–814.

Meehan, K. (2019) "Water justice and the law in Latin America," *Latin American Research Review*, 54 (2), pp. 517–523. doi.org/10.25222/larr.461

Meehan, K.M. and Moore, A.W. (2014) 'Downspout politics, upstream conflict: formalizing rainwater harvesting in the United States,' *Water International*, 39 (4), pp. 417–430.

Olmos Giupponi, M.B. and Paz, M.C. (2015) 'The implementation of the human right to water in Argentina and Colombia,' *Anuario Mexicano de Derecho Internacional*, 15 (1), pp. 323–352.

Rabasa, A., Velasco, A. and Martínez, X. (2014) *La Instrumentación del Derecho Constitucional al Agua en México: recomendaciones para su Regulación*. Mexico City, Mexico: Centro Mexicano de Derecho Ambiental and the Environmental Law Institute.

Roa-García, M.C., Urteaga-Crovetto, P., and Bustamante-Zenteno, R. (2017) 'Water law in the Andes: a promising precedent for challenging neoliberalism,' *Geoforum*, 64 (1), pp. 270–280.

Seemann, M. (2016) *Water security, justice, and the politics of water rights in Peru and Bolivia*. New York: Palgrave Macmillan.

United Nations Human Rights Council (2017) *Report of the Special Rapporteur on the human right to safe drinking water and sanitation on his mission to Mexico* (A/HRC/36/45/Add.2). Available at: http://ap.ohchr.org/documents/dpage_e.aspx?si=A/HRC/36/45/Add.2

United Nations Office of the High Commissioner for Human Rights (2017, May 12) 'UN expert calls on Mexico to urgently expand and improve water and sanitation provision for all.' Available at: www.ohchr.org/EN/NewsEvents/Pages/DisplayNews.aspx?NewsID=21618&LangID=E

Wilder, M. (2008) 'Equity and water in Mexico's changing institutional landscape,' in Whiteley, J.M., Ingram, H., and Perry, R.W. (eds) *Water, place, and equity*. Cambridge: The MIT Press, pp. 95–116.

Wilder, M. (2010) 'Water governance in Mexico: Political and economic apertures and a shifting state-citizen relationship,' *Ecology and Society*, 15, (2), pp. 22. Available at: www.ecologyandsociety.org/vol15/iss2/art22

Wright, M.W. (2017) 'Epistemological ignorances and fighting for the disappeared: lessons from Mexico,' *Antipode*, 49 (1), pp. 249–269.

Wright, M.W. (2018) 'Against the evils of democracy: fighting forced disappearance and neoliberal terror in Mexico,' *Annals of the American Association of Geographers*, 108 (2), pp. 327–336.

4 Turning to traditions

Three cultural-religious articulations of fresh waters' value(s) in contemporary governance frameworks

Christiana Zenner

Introduction: cultural-religious traditions and plural waters

Fresh waters are dynamic, intersectional issues: they mediate relationships at every level of scale. Decisions and actions affecting those relationships constitute major twenty-first-century issues for both governance and ethics in ways that require foregrounding three baseline moral assumptions. First, fresh waters are morally intriguing because while they are sui generis (non-substitutable) and sine qua non (essential for life), fresh waters are neither singular nor uniform; they are plural and multiform. Their native watershed dynamics, institutionalized and technologized flows, incentivized uses, uneven distributions, and myriad valuations appear in customary cultural forms as well as neoliberal frames. Second, as human geographers have pointed out, water is not merely a physical entity but is also always interwoven with human arrangements of political economy, social structure, and power. As Linton and Budds summarize:

> The hydrosocial cycle is based on the concept of the hydrologic cycle, but modifies it in important ways. While the hydrologic cycle has the effect of separating water from its social context, the hydrosocial cycle deliberately attends to water's social and political nature. . . . [It is] a socio-natural process by which water and society make and remake each other over space and time.
> (Linton & Budds, 2014; see also Boelens, Vos, & Perrault, 2016)

Third, cultural-religious moral frameworks are crucial sites for exploring and defining what sorts of things fresh waters are understood to be in ways that may both align with, and contrast to, mainstream human rights discourse. Interdisciplinary approaches in religious studies and anthropology can (at best) illuminate how waters carry apparent linkages "between the social and the cosmological," as Fabienne Wateau has summarized, including where "social rationales [for water access and distribution] can be stronger than economic ones . . . The problem that must be faced today is the way in which local and social, cosmological and symbolic rationales clash with global and economic, technical and efficiency rationales" (Wateau, 2011, p. 259, 263–264).

Intellectual genealogies of western scholarship are colonialist and triumphalist in origin (and often in an ongoing expression), and the study of religious ethics is no exception. At best, however, it can also facilitate analyses of moral reasoning that do not shy away from normativity and may be of service in framing scholarly work that is rigorously descriptive but also has a prescriptive element—in this case, an orientation toward expanding conceptions of fresh waters' ontologies and ethical obligations that follow. The project of this chapter is to explore what happens when such religious or cultural conditions are mobilized within the dominant framework of Western liberal governance discourses, focusing on three cases: *mni wiconi* at Standing Rock, the Whanganui in Aotearoa/New Zealand, and the official teachings of the Catholic Church.

Religious conceptualizations, especially place-based Indigenous value systems and ritualized cultural formulations, tend to align relatively well with the idea of hydrosociality and can demonstrate how hydrosocial constructs bear real, embodied, and other forms of moral consequences (Drew, 2016; Phare & Mulligan, 2015). The ongoing and increasing visibility of diverse Indigenous voices as cultural-religious authorities in ethical objection to, and legal unity against, the onslaught of industrial Western values in the age of digitally mediated activism is occurring alongside a widespread scholarly turn to the value of Indigenous communities' embedded ecological knowledge, social structures, and self-determination. However, such a turn is fraught in several ways. Specifically, the (re)current turn to Indigenous ecological wisdom is a major epistemic and ethical issue in a colonially-shaped, capitalist, information-driven age where knowledge production has often subjugated non-white populations and those who have been historically marginalized or overwritten in terms of territory, bodily autonomy, language, and cultural values (Norman, 2017; Snelgrove, Dhamoon, & Corntassel, 2014; TallBear, 2014; Tuck & Yang, 2012). Dominant academic practice—and those who embody its privileges—has usurped the voices and experiences of marginalized groups and individuals through a lack of accountability to the communities from whom knowledge is gleaned, historically and into the present day. These conceptual and embodied dangers of colonial reinscription apply to white, Western scholars like myself as well as institutions like the Catholic Church—no matter how well intended the exhortation to expand one's epistemology.

Indigenous peoples, especially in this instance water advocates and protectors, are their own authorities. Western governance paradigms and mainstream discourses of value would do well to attend to these epistemologies, value frameworks, and ethical stipulations (Christian & Wong, 2017; Zenner, 2019). My attempt is to foreground respectfully and accurately the self-articulations of religious or cultural formulations that challenge normative presumptions of Western, neoliberal governance frameworks, including rights paradigms; to note how those systems of knowledge are mobilized in the context of legal strategies; and to braid together the similarities and differences in these deployments of western political-economic-legal notions. I hope that this serves to amplify (but by no means define or exhaust) the insights and moral claims of these communities of knowledge while recognizing that their practitioners and representatives also always speak best for themselves.

The following sections analyze context and discourse from: (1) claims to territorial sovereignty and *mni wiconi* at Standing Rock (United States); (2) the claims of personhood for the Whanganui in Aotearoa (New Zealand); (3) an interlude on the philosophical frameworks that undergird the United Nations' articulation of a Human Right to Water, in contrast to the Declaration of the Rights of Mother Earth; and (4) the Catholic Church's emphatic endorsement of the HRW, with its recent and potentially highly problematic corollary exhortations about the value of Indigenous knowledge.

Mni wiconi: Standing Rock

While media publicity surrounding the actions of the Water Protectors at Standing Rock effloresced after 2014, the roots of history run much deeper into legacies of settler colonialism and related government infrastructure projects. Following a half-century in which Army Corps of Engineers projects dammed and diverted various water sources, the Lakota Sioux of the Standing Rock reservation in North Dakota advocated for sufficient access to clean, fresh water for their lands and people. In November 2004, a Senate hearing convened by the Committee on Indian Affairs was held on objections pertaining to Army Corps of Engineers' diversion of portions of Missouri River tributaries to facilitate barge commerce downstream. Standing Rock Sioux chairman Charles Murphy at that time presented evidence that the Corps' actions had negatively affected the water supply of the reservation:

> We don't have the water to provide for our people. One year ago . . . we had approximately 10,000 people without water. These were Indian and non-Indian people within our reservation of 2.3 million acres [and 18,000 people]. . . . Senator, we have a major issue out there with the management of the Missouri River situation.
> (US Senate Committee of Indian Affairs, 2004, p. 3–4)

Tim Johnson, Senator from North Dakota, added that the situation is "particularly disconcerting given the treaties that bind the Federal Government's responsibility to our tribes in North and South Dakota" (US Senate Committee of Indian Affairs, 2004, p. 5). The relevance of concerns over access to water as a justice issue, a structural problem, and a militaristic connection was clear to Senator Daniel Inouye, chair of the Committee on Indian Affairs, who noted the parallels between the Army Corps of Engineers' actions affecting the Standing Rock Sioux, on the one hand, and the US military's reception in Iraq, on the other hand—both of which led to warranted "faces of anger . . . in our hearts we knew that there were many causes for this, [including] that we did not have plans to repair the damaged water systems and the damaged sewer systems" (US Senate Committee of Indian Affairs, 2004, p. 22). Inouye here reveals the militaristic-colonial complex that mediates access to water between and within territories.

Given these historical precursors and structural tendencies, it was perhaps not surprising that the Army Corps of Engineers fast-tracked the Dakota Access Pipeline (DAPL) through the Standing Rock reservation to transport fossil fuels from the Bakken oil shale, crossing and traveling under the Missouri River. Energy Transfer Partners, the operating company for DAPL, asserted that it has followed due procedures in soliciting input from the Sioux. Representatives from the reservation disagreed and filed a lawsuit claiming that due process was violated.

The conflict at Standing Rock embodies a tangle of disenfranchisements over water, land, sovereignty, and economic benefit that has unique features even as it is also simultaneously emblematic of broader structural patterns of disenfranchisement and settler colonialist legacies wrought upon the lives, cultures, and lands of native peoples in the United States (Dunbar-Ortiz, 2014). The embodied burdens are real, and the pipeline will disproportionately affect vulnerable Native American populations, as research by geographers Jennifer Veilleux and Candace Landry indicates (Veilleux, 2016). Thus, #NODAPL actions are predicated on two major, interwoven ethical claims: first, Indigenous sovereignty should outweigh for-profit/neocolonial and government-backed expansion of extractive industry infrastructure; second, *mni wiconi*—water is life, and as such the risk of its contamination (for both the earth and marginalized bodies) is not acceptable.

Chairman David Archambault II specified how this is a "familiar story in Indian Country," adding: "Now the Corps is taking our clean water and sacred places by approving this river crossing. . . . Protecting water and our sacred places has always been at the center of our cause" (Archambault, 2016). As the American Civil Liberties Union (ACLU) points out, the actions of #NODAPL protestors drew major intersectional support:

> More than 200 tribes and several thousand indigenous people from across the country have gathered in North Dakota to protest the Dakota Access Pipeline. The protesters are defending the land and water using little more than the right to assemble and speak freely—a protection afforded by the U.S. Constitution. In response to the protests, North Dakota's government suppressed free speech and militarized its policing by declaring a state of emergency, setting up a highway roadblock, and calling out the National Guard.
>
> (ACLU, n.d.)

Brenda White Bull, a Lakota woman and US military veteran, stated simply: "The highest weapon of them all is prayer. . . . The world is watching. Our ancestors are watching. . . . We are fighting for the human race" (cited in Erdrich, 2016). Many US military veterans journeyed to North Dakota to protect the native bodies who were, in turn, protecting the water, including the Native American veterans who "serve in the US military at a higher rate than any other ethnic group" (Erdrich, 2016). Tribal chairman David Archambault II addressed veterans with gratitude: "What you are doing is precious to us. I can't describe the feelings that move over me. It is *wakan*, sacred. You are all sacred" (Erdrich, 2016).

The legal efficacy of appeals to tribal sovereignty has been desultory and muddled in the case of DAPL. There was some optimism when in December 2016 then-President Obama did not grant a last easement to facilitate pipeline construction under Lake Oahe and, instead, required a further environmental impact assessment. But in the age of Trump, the easement was quickly (re)granted, a move that drew an immediate legal challenge from the Sioux.

From the perspective of international institutions, Robison et al. point out that the UN Permanent Forum demonstrated how "the United States and its political subdivisions had transgressed UNDRIP [United Nations Declaration on the Rights of Indigenous Peoples] repeatedly in their dealings with the people of the Great Sioux Nation over DAPL" (Robison et al., 2018, p. 875). They add that "DAPL illuminates the historical and contemporary phenomenon . . . [of] Indigenous Peoples' struggles for justice in relation to the basis of life—water" (Ibid., 876). For Louise Erdrich the enduring image is of a person, sprayed by water cannons deployed by the government, not protesting but protecting values and land—"covered in ice and praying, [illustrating] the resolve that comes from a philosophy based on generosity of spirit" (Erdrich, 2016).

Indeed, while tribal sovereignty is the basis of the legal challenges to DAPL, it is ideas of the sacred that permeate water protectors' actions. Prominent slogans include "we are water," "water is sacred," and *mni wiconi*—water is life. All of these phrases—and their philosophical underpinnings—are potential portals to twenty-first century appeals to justice in the face of historic and ongoing territorial and cultural disenfranchisement. *Mni wiconi* summons the idea of the sacredness of waters that sustain human beings, communities, and more-than-human ecosystems, moving well beyond languages of human rights (as discussed in the following sections). These phrases appear on posters both handmade and digital, on the front lines of protests and solidarity/fund-raising concerts, and on the home pages of the Indigenous Environmental Network and the Standing Rock Sioux. As the website of the Standing Rock Sioux phrases it: "In honor of future generations, we fight this pipeline to protect our water, our sacred places, and all living beings" (Standing Rock Sioux, 2017).

Māori advocacy for the Whanganui of Aotearoa

Thousands of miles across the Pacific Ocean, the Māori of Aotearoa, New Zealand, have also engaged legal mechanisms in a settler-colonial governance milieu in order to claim cultural and communitarian water rights in the face of the impoundment and privatization of water resources. The Māori have since 1840 been a party to the Treaty of Waitangi, in which the colonial government of New Zealand acknowledged the community's knowledge and claim to ownership of land and derivatively (though not explicitly) to "rights and interests" in waters. More recently, when the New Zealand government "planned to sell shares of publicly owned hydroelectricity companies (and thus their vast water allocations)," Maori communities protested that "living water" is a "being whose spirit remains present at the spring and at specific places along the stream" and is "a generative 'life essence' of people and place," and even that the river "is the people themselves" (Strang, 2014, 124–128). Veronica Strang has depicted how the government persisted in its determination to sell off shares of the hydroelectric resource, leading to a lawsuit that went to the Supreme Court, which ruled that a mixed-ownership regime (between Maori and the state) was necessary. She argued in 2014 that a positive result is the expansion of national laws "to accommodate explicit notions of long-term affective attachments to place; the ways that social and cultural identity are encoded in material environments; and ideas about living water and spiritual connections" (Strang, 2014, p. 122).

Following this development, it was major international news in March 2017 when the river was granted status as a legal entity equivalent to a person. The *Guardian* described it as a "world-first" that the river "has been granted the same legal rights as a human being" (Roy, 2017). In truth, it is probably more accurate to say that this is a "world-first" since the domination of cultures and territories by European colonial powers and subsequent neoliberal forms of life, but the framing of the sentence itself suggests the totalizing power and assumptions of liberal Western legal and political categories in the present day.

As lead tribal negotiator Gerrard Albert noted in 2017, the grounds of the moral claim to the river's legal status are ancestral and cultural: "The reason we have taken this approach is because we consider the river an ancestor and always have," he said. But insofar as this concept of nonhuman entities lacks cognates in Western legal and economic frameworks, the choice of rights language based on the entirety of the value of the river was intentional and strategic. In Albert's words: "We have fought to find an approximation in law so that all others can understand that from our perspective treating the river as a living entity is the correct way to approach it, as in indivisible whole, instead of the traditional model for the last 100 years of treating it from a perspective of ownership and management" (Roy, 2017). Here, ancestral cultural claims aligned with ongoing treaty litigations to culminate in what Liz Charpleix calls "a landmark political decision recognizing the legal personhood of a river." She sees the 2017 decision as an instance that "provides insights into how legal pluralism may evolve" between place-based conceptions of ecological integrity and dominant legal frameworks. She suggests that this and other "less anthropocentric" conceptions of nature could begin to emerge even within dominant legal frameworks (Charpleix, 2018). Indeed, wrote Jacinta Ruru in an anticipatory way, the moral claim is about "a new angle [in western legal frameworks] that centres on an Indigenous right to water *for identity*. The wellbeing of the Maori depends on the wellbeing of water" (Ruru, 2012, p. 120).

Particularly at issue is the question of what bases for assigning a legally recognizable ontology are admissible in contemporary liberal Western frameworks. The case of the Whanganui specifically provides one example of how biophysical entities that long preceded *Homo sapiens* might be regarded not just as relics of minority cultural cosmologies but also as functional, legal persons (see Youatt, 2017). More pointedly, we might observe how, in the present era of late capitalism, few denizens of dominant cultures question whether a corporation is assumed to be a legal person—yet the assignation of personhood to a river in 2017 struck many as odd. The conditions leading to such category disjuncture is a moral inconsistency well worth pursuing.

Declarations and their discontents

The idea of a Human Right to Water (HRW) has since 2010 been enshrined in United Nations formulations and appears in the constitutions of several nations (such as South Africa), as well as subsidiary political bodies (such as the state of California). The language of the human right to water has been described as useful, a necessary advance in stipulating moral parameters to legal paradigms, as a precondition for the achievement of all other rights, and—along with the right to sanitation—as a massive step forward in the pursuit of gender equity. Yet the idea of the human right to water has also been criticized for being difficult to implement; for serving to facilitate commoditization and privatization as modes of achieving the right to water; and, perhaps most important, for being anthropocentric in ways that reflect a Western philosophical and property orientation that insufficiently recognizes the interdependencies and autonomies that fresh waters facilitate, carry, and undergird. Even from within Western discourses, there are evident shortcomings of the HRW framework, for example insofar as it fails to acknowledge linkages between domestic and ecological functions (Angel & Loftus, 2019; Mirosa & Harris, 2012).

As is evident by its omission from the preceding accounts, the language of HRW does not appear centrally in the advocacy about Standing Rock or Māori articulations of the Whanganui's ontology, perhaps because those contextually- and culturally-specific notions of water exceed Western philosophical anthropocentrism. For the Māori and for the Sioux, as well as for many other Indigenous religious and cultural traditions, it is precisely the opposite ethical obligation that is foregrounded: human beings have an ethical obligation to protect the waters and, in the case of the Whanganui, the river has agency and identity in itself.

The philosophical disjuncture is nicely drawn out by a comparison of the language of HRW with the language in the Declaration of the Rights of Mother Earth (World People's Conference on Climate Change and the Rights of Mother Earth, 2010). The 2010 "Declaration of the Rights of Mother Earth" resulted from the massive World People's Conference on Climate Change in Bolivia. The Declaration identifies certain entitlements that are due to the Earth—notably imaged here as Mother—"without distinction of any kind, such as may be made between organic and inorganic beings, species, origin, use to human beings, or any other status." The document further holds that Mother Earth has a right "to regenerate its biocapacity and to continue its vital cycles and processes free from human disruptions," which prominently includes "the right to water as a source of life" (Ibid.)

The Declaration foregrounds how Indigenous accounts of fresh waters are often non-anthropocentric (not focused primarily or exclusively on humans) and may be resolutely biocentric (valuing all biotic life or the biosphere) or ecocentric (valuing the earth and its integrating systems, including but not limited to living beings). More specific implications follow when the philosophical orientation is put into the context of legal vernaculars. For example: how far can or should the sphere of rights entitlements extend—for example, might other animals or ecosystems have rights (in this case, to the integrity of waters)? Even more strongly, might water "itself"—however understood—have a right to exist, for example, in an uninterrupted, undredged state? In such frames, the moral anthropocentrism of HRW discourse is a woefully incomplete formulation of the ways that human beings and waters relate to one another. Yet despite these objections, the anthropocentric HRW has received significant attention within international governance and dominant, Western scholarly and activist discourses—including the emergent ecological doctrines of the Catholic Church, specifically those that focus on fresh water.

The Catholic Church: universality without geography

The Catholic Church has a decisive characteristic of being a centralized authority with a clear leader and a century of aggregated, official moral teaching on matters of social concern. It has also, for most of its modern history, had the dubious distinction of being explicitly and radically anthropocentric (and also androcentric) such that, in 1990, Columban missionary and priest Fr. Sean McDonagh indicted the magisterial Catholic Church's failure to address the intersection of environmental and social degradations. His frustration was the well-honed result of years spent working alongside people in territories whose lives were constrained by social and ecological injustices, many of which flowed from the dynamics of colonialism and economic globalization. "The Church has been slow to recognize the gravity of the ecological problems of the earth," he wrote, condemning as well the problematic endurance of "domination theology" with its universalizing "anthropocentric bias" (McDonagh, 1990, 175–176).

While there is more to the story of the Catholic Church's ecological turn, what is noteworthy for the purposes of this chapter is how the Church has deployed the language of rights, particularly the human right to water (Zenner, 2018a). On a governance level, this is not surprising, since the Catholic Church has been a steady presence as an observer state at the United Nations and was an early advocate at the UN for the idea of the human right to water as formulated in 2002. And in genres holding among the highest levels of authority in Catholic documentary tradition, popes have underscored the importance of enshrining a human right to water in international law (Zenner, 2018b). Such papal statements, of course, tend to be exhortatory in tone and are not so much legal documents as pastoral and moral reflections meant to guide the ethical awareness and actions of a specific (though numerically significant) religious group.

For the purposes of this essay, there are instructive differences between the previous two cases of Standing Rock and the Whanganui, on the one hand, and the Catholic Church, on the other. Indigenous advocacies at Standing Rock and for the Whanganui have selectively mobilized Western philosophical and legal frameworks towards particular, place-based outcomes while centralizing their own ontological-ethical understandings of fresh water. The religious-ethical emphasis on the HRW by the Catholic Church, by contrast, appeals to dominant Western philosophical/anthropocentric notions through the vehicle of HRW and the UN, is not at all placed-based, and, thus, makes universalizing moral claims without attention to particular social and ecological contexts.

Some water justice advocates might be quick to respond that there are benefits to the Catholic Church's approach, which holds that the right to water is understood as fundamental to the achievement of all other rights and that it is the responsibility of states to provide water for their inhabitants in an equitable fashion, regardless of ability to pay. Specifically, the visibility of a prominent global figure making pronouncements about fresh water and human rights should not be understated, as when he addressed a 2017 summit at the Vatican on the HRW:

> It is essential that each state act as a guarantor of universal access to safe and clean water. . . . With the 'little' we have, we will be helping to make our common home a more liveable and fraternal place, better cared for, where none are rejected or excluded, but all enjoy the goods needed to live and to grow in dignity.
>
> (Francis, 2017)

But in addition to the differences in terms of anthropocentrism, the Church's endorsement of the HRW differs from the advocacy of the Whanganui or Standing Rock in several key ways. For one thing, it is not born out of a specific, place-based reality with the explicit goal of obtaining more capacious hydrosocial accommodations in a particular nation-state. For another, it is disseminated by a globally-dominant religious institution not standardly regarded as ecologically activist (though, with Francis, this may be shifting). Third, it replicates a center-periphery power-and-knowledge dialectic between Western philosophical and theological concepts and institutions, on the one hand, and Indigenous traditions and representatives, on the other.

In his ecological document *Laudato Si'* (2015), Pope Francis noted several times that Indigenous communities' value systems deserve respect and underscored the importance of their moral authority on projects affecting their ancestral lands. To put it mildly, such a turn is new for the Catholic Church, and it is by no means yet clear whether this is a principle that will be positive or pernicious (Zenner Peppard, 2018). Positively, the papal turn to Indigenous value systems and wisdom could serve to support and make legible a range of Indigenous ecological activisms on terms that dominant groups (who often oppose those activisms) would recognize. If so, perhaps this would provide a type of institutional protection for environmental activists who are persecuted, especially within historically Catholic nations; that would be a salutary outcome. But it is by no means clear that the pope's appeal to Indigenous knowledge will be taken up in this way in practice. It is also not clear that the papal appeal will be explored responsibly in practice by Western theologians and scholars, not to mention the institutional Catholic Church itself, given the legacies of colonialism and domination that have long been promulgated and justified within those very structures of academic knowledge production and colonial, missionary institutional expansion. For a centralized institution like the Catholic Church to undertake a dissolution of epistemological and ethical hierarchy between center and periphery when it comes to Indigenous ecological values would be very radical and welcome indeed, but this is not historically demonstrated in an institution that has consistently reinscribed exactly the opposite (Zenner, 2019).

Conclusion

This essay has described several different modes through which religious-cultural articulations of fresh waters' ontologies have been deployed in recent years. Each of the formulations described in this chapter is shaped variously within, by, against, and beyond late capitalist formulations of property and commodity logic as well as a range of governance paradigms. The advocacy surrounding Standing Rock (the United States) and the Whanganui (New Zealand) are clearly articulations of diverse Indigenous values and more-than-anthropocentric moral ontologies for water against a backdrop of specific settler-colonial legal paradigms and political contexts. By contrast, the generalized anthropocentric Catholic conceptual formulation of fresh waters' significance centralizes the anthropocentric HRW as enshrined in mainstream international governance discourses, without attention to particular contexts.

Through these cases, this chapter has pointed out how fresh waters refract multiple notions of ontology, which can be mediated through the language of legality as well as the sacred, and which can be strategically deployed against or in alignment with specific powerful governance frameworks in the present day. It has warned against some of the potential perils of a neo-colonial turn to reify Indigenous knowledge and values. As such, this chapter has also demonstrated how humanistic discourses can be vital conversation partners toward more capacious moral ontologies for fresh waters, by rendering visible some of the ways that histories and cultures nuance, refute, or skew away from dominant ideologies and historically-shaped political-economic regimes of late capitalism—the last of which, it must be said, can be as frequently and fervently espoused as any religious dogma.

References

ACLU of North Dakota (no date) 'Stop government suppression of the right to protest in North Dakota.' Available at: https://action.aclu.org/secure/nd-standing-rock-sioux-tribe

Angel, J. and Loftus, A. (2019) "With-against-and-beyond the human right to water," *Geoforum*, 98, January, pp. 206–213. doi:10.1016/j.geoforum.2017.5.002

Archambault, D. (2016) 'Taking a stand at Standing Rock,' *The New York Times*, 24 August. Available at: www.nytimes.com/2016/08/25/opinion/taking-a-stand-at-standing-rock.html

Boelens, R., Vos, J., and Perrault, T. (2018) 'Introduction: the multiple challenges and layers of water justice struggles,' in Boelens, R., Perrault, T., and Vos, J. (eds) *Water justice*. Cambridge: Cambridge University Press.

Charpleix, L. (2018) 'The Whanganui river as Te Awe Apua: place-based law in a legally pluralistic society,' *The Geographical Journal*, 184 (1), pp. 19–31.

Christian, D. and Wong, R. (eds) (2017) *Downstream: reimagining water*. Waterloo, Canada: Wilfred Laurier University Press.

Drew, G.L. (2016) 'Beyond contradiction: sacred-profane waters and the dialectics of everyday religion,' *Himalaya*, 36 (2), pp. 70–78.

Dunbar-Ortiz, R. (2014) *An Indigenous people's history of the United States*. Boston, MA: Beacon Press.

Erdrich, L. (2016) 'Holy rage: lessons from Standing Rock,' *The New Yorker*, 22 December. Available at: www.newyorker.com/news/news-desk/holy-rage-lessons-from-standing-rock

Francis (2015) '"Laudato si": on care for our common home.' Available at: http://w2.vatican.va/content/francesco/en/encyclicals/documents/papa-francesco_20150524_enciclica-laudato-si.html

Linton, A. and Budds, J. (2014) 'The hydrosocial cycle: defining and mobilizing a relational-dialectic approach to water,' *Geoforum*, 57, pp. 170–180.

Francis (2017) 'The pope speaks at seminar on the human right to water.' Available at: https://press.vatican.va/content/salastampa/en/bollettino/pubblico/2017/02/24/170224a.html

McDonagh, S. (1990) *The greening of the Church*. Maryknoll, NY: Orbis Books.

Mirosa, O. and Harris, L. (2012) 'Human right to water: contemporary challenges and contours of a global debate,' *Antipode*, 44 (3), pp. 932–949. Available at: doi:10.1111/j.1467-8330.2011.00929.x

Norman, E. (2017) 'Standing up for inherent rights: the role of indigenous-led activism in protecting sacred waters and ways of life,' *Society and Natural Resources*, 30 (4), pp. 537–553. Available at: doi:10.1080/08941920.2016.1274459

Phare, M. and Mulligan, B. (2015) 'Water governance: restoring sustainable use through indigenous values,' in Davidson, S., Linton, J., and Mabee, W. (eds) *Water as a social opportunity*. Kingston, ON: Queens University Policy Studies.

Robison, J., Cosens, B., Jackson, S., Leonard, K., and McCool, D. (2018) "Indigenous water justice," *Lewis and Clark Law Review*, 22 (3), pp. 873–953.

Roy, E.A. (2017) 'New Zealand river granted same legal rights as human being,' *The Guardian*, 16 March. Available at: www.theguardian.com/world/2017/mar/16/new-zealand-river-granted-same-legal-rights-as-human-being

Ruru, J. (2012) 'The right to water as the right to identity: legal struggles of Indigenous peoples of Aotearoa New Zealand,' in Sultana, F. and Loftus, A. (eds), *The right to water: politics, governance, and social struggles*. New York: Routledge.

Snelgrove, C., Dhamoon, R.K., and Corntassel, J. (2014) 'Unsettling settler colonialism: the discourse and politics of settlers, and solidarity with Indigenous nations,' *Decolonization: Indigeneity, Education, and Society*, 3 (2), pp. 1–32.

Standing Rock Sioux (2017) 'Stand with Standing Rock.' Available at standwithstandingrock.net

Strang, V. (2014) "The Taniwha and the Crown: defending water rights in Aotearoa/New Zealand." *WIREs Water*, 1, pp. 121–131. doi:10.1002/wat2.1002

TallBear, K. (2014) 'Standing with and speaking as faith: a feministindigenous approach to inquiry,' *Journal of Research Practice*, 10 (2), Article N17. Available at: http://jrp.icaap.org/index.php/jrp/article/view/405/371

Tuck, E. and Yang, W. (2012) "Decolonization is not a metaphor," *Decolonization: Indigeneity, Education & Society*, 1 (1), pp.1–40.

United States Senate Committee of Indian Affairs (2004) 'Water problems on the Standing Rock Sioux reservation.' Available at: www.gpo.gov/fdsys/pkg/CHRG-108shrg97093/pdf/CHRG-108shrg97093.pdf

Veilleux, J. (2016) 'Income maps of the Native Americans living in the Missouri river basin,' *The Way of Water*, 19 December. Available at: http://jveilleux.blogspot.com/2016/12/income-maps-of-native-americans-living.html?m=1

Wateau, F. (2011) 'Water, societies, and sustainability: a few anthropological examples of non-market water values,' *Policy and Society* 30, pp. 257–265. Available at: doi:10.1016/j.polsoc.2011.10.004

World People's Conference on Climate Change and the Rights of Mother Earth (2010) 'Declaration of the Rights of Mother Earth.' Available at: https://pwccc.wordpress.com/programa

Youatt, R. (2017) 'Personhood and the rights of nature: the new subjects of contemporary Earth politics,' *International Political Sociology*, 11 (1), pp. 39–54.

Zenner, C. (2018a) '"Laudato si" and Standing Rock: water justice and Indigenous ecological knowledge,' in Deane-Drummond, C. and Artinian-Kaiser, R. (eds) *Theology and ecology across the disciplines: on care for our common home*. New York: Bloomsbury.

Zenner, C. (2018b) *Just water: theology, ethics, and global fresh water crises*. Maryknoll, NY: Orbis Books.

Zenner, C. (2019) "Valuing fresh waters," *WIREs-Water*, 6 (3), pp. 6:e1343. doi:10.1002/wat2.134

Zenner Peppard, C. (2018) 'Commentary on Laudato si',' in Himes, K. (ed), *Modern Catholic social teaching: commentaries and interpretations*. Washington, DC: Georgetown University Press.

5 The right to bring waters into being

Jamie Linton

Introduction

In this chapter, we make an effort to reconcile the concept of the right to water with recent academic work that brings into question the political dimensions of water's ontological status. A relational-dialectical approach to water has already produced the argument that water is no one thing and that different waters are instantiated in different circumstances that could be described as both social and natural (Linton, 2010). Taking this argument a step further, it was suggested that people could intentionally produce a particular type or version of water by changing these circumstances: For example, what might be described as the public water flowing through a publicly-owned and managed water system *is* constitutionally different from the commercial water distributed and sold in individual bottles (ibid., Chapter 12). Not only are these waters constitutionally different, but also, they produce different social and natural effects. The relations that underpin these arguments have been formalized in a concept known as the hydrosocial cycle, defined as "a socio-natural process by which water and society make and remake each other over space and time. . . [emphasizing that] the hydrosocial cycle comprises a process of co-constitution as well as material circulation" (Linton and Budds, 2014).

These ideas complement recent work that focuses directly on "multiple ontologies of water" (Yates, Harris and Wilson, 2017). Ontology is the branch of philosophy that treats of fundamental questions being and becoming. As will be discussed, considering multiple ontologies of water follows a trend most notably advanced in anthropology and in science studies (but including geography[1]), described as "the ontological turn." To speak of "ontologies" in the plural is to recognize that the very being of something is different, depending on the circumstances of its becoming. As STS scholar Annemarie Mol has argued: "Now the word needs to go in the plural. For. . . if reality is *done*, if it is historically, culturally and materially *located*, then it is also *multiple*. Realities have become multiple" (Mol, 1999 p. 75).

Not surprisingly, questions surrounding the meaning and practice of politics in a world where "realities have become multiple" are open-ended. Here, we consider the question of who has the right, and the power, to bring a particular water into being – or even to the table – in the context of water governance. This is a pertinent question in places where "modern water" (Linton, 2010) is confronted with alternative waters, such as in parts of Canada where First Nations peoples' relations with and ontologies of water come into play. As several researchers have pointed out, the ontological basis of First Nations peoples' knowledge claims has seldom been taken seriously in the context of land claims and co-management strategies involving water and other resources (Nadasdy, 2003; Nadasdy, 2007; Blaser, 2016; Yates, Harris & Wilson, 2017; Wilson & Inkster, 2018). At best, the testimonies and practices of First Nations peoples that suggest alternative ontologies are treated as "cultural constructions" (Nadasdy, 2007, p. 26). In this chapter, we suggest that the concept of rights might be considered as a strategy towards taking seriously the fact of ontological difference where it applies to water.

We begin by describing the ontological turn and the notion of multiple ontologies of water. This is followed by a discussion of the project of "political ontology" that has been advanced as a way of validating the ontological claims of indigenous peoples and affording them a measure of power or influence in projects of resource management – and the extension of this argument to water management. We then turn to the question of the right to water and ask how it might be possible to extend the argument for water rights to apply to situations where multiple ontologies of water prevail. We argue this complements the political ontology project and has the potential of strengthening the power of indigenous peoples and others in their struggles against water-resources development discourse and projects.

The ontological turn and multiple ontologies of water

"The ontological turn" describes a theoretical trend developed most notably in Science and Technology Studies (STS) and anthropology that upsets the modern presupposition that there exists a single, objective reality.[2] That modern ontological assumptions are undergoing a crisis is illustrated by dissatisfaction with what STS scholar John Law describes as "the one-world world" (Law, 2011). This is the idea, traceable to Descartes, which holds there is a single reality which is knowable only by means of representation; and that while our representations, our discourses, our social constructions, and our cultural interpretations of the world may differ, they are but various partial and imperfect renderings of the ultimate underlying reality of things. One important political implication of this way of thinking is that there must be some renderings of reality that are more accurate or more reliable than others – that get closer to the real reality of things – and serve as a basis for diminishing, discrediting or dismissing those of others. The ontological turn is arguably having the effect of rendering such dismissiveness less tenable. In the current climate of crisis and dissatisfaction with modern, Cartesian ontology, what appear to be the (alternate) realities of others – particularly those who have been colonized and marginalized in the march of modernity – become interesting, appealing and insightful, whereas in the one-world world, they could be nothing but quaint, superstitious or absurd.

If there is a single theme that is common to proponents of the ontological turn, it is that reality does not pre-exist its becoming but is always in-the-making as an effect or product of the assemblages, enactments, worldings or performances which bring things into being. To cite just a few examples: The anthropologist Mario Blaser writes "ontology is a way of worlding, a form of enacting a reality" (Blaser, 2013 p. 551). For Annemarie Mol, "reality is done" and, thus, "reality is historically, culturally and materially located" (Mol, 1999 pp. 87, 75). And John Law writes, "*Different realities are enacted in different practices*, and this is a chronic condition" (Law, 2011 p. 5). Reality, in other words, isn't simply out there waiting to be discovered and observed, but it is always occurring as a consequence of different forms of practice and engagement, including discovery and observation.

Turning to water, we understand "water" as a stable entity – represented as H_2O – that remains the same, no matter the circumstances. When liquid water turns to ice we say that water freezes and when it turns to vapor we say that water vaporizes – because the essence of water is that to which all these things can be reduced. We can contrast this with the classical approach, which was present in Western thinking as late as the late 18th century when Lavoisier famously made it known that water was a compound of hydrogen and oxygen. As per Aristotle's theory of the four elements, ice was not water but a form of earth. And water vapor was not water but rather a form of air. Aristotle certainly knew that if you left a bucket of liquid water out overnight and it froze, it would turn to ice. But for him, the important ontological consideration was not to focus on the common identity to which both liquid water and ice could be reduced; the important thing was the way the elements actually changed in relation to the circumstances. Thus, for Aristotle, when (the quality of) coldness was replaced by heat, water became air, and when moistness was replaced by dryness, it became ice. In other words, the elements underwent transubstantiation – they were, in Aristotle's words, "transformed into one another" by means of undergoing changes in quality (Sambursky, 1962 p. 32).

To bring people into the picture, in his Paris laboratory, Lavoisier passed a quantity of water through an apparatus that broke it into its two elementary components (which he later named hydrogen and oxygen), thus finally proving, or devising a way of saying, that water was not an element after all (Dear, 2006 p. 79). But more importantly for our purposes, this was the means by which Lavoisier assisted in bringing what we could call "modern water"[3] into the world, through a scientific experiment. We might say that H_2O was always present as a (latent) possibility but that it was worlded, enacted or realized in this scientific practice. And once realized, made manifest and represented, modern water became an important (constitutive) part of our world, reproducing itself through further epistemic and engineering practices, discourses, infrastructures, laws, regulations, textbooks and water quality standards. Historically, modern water overturned worlds of myriad different "waters" that were present in western Europe and beyond, realized and sustained in a wide variety of practices (Hamlin, 2000). Today modern water endures as the basis for "water resources" discourse and development, despite widespread crises, globally and locally (Linton, 2010). Indeed, it represents the hegemonic "ontology" of water against which "multiple ontologies of water" have to contend.

In recent years, scholars have begun to think about water in light of the ontological turn. In 2012, Jessica Barnes and Samer Alatout coordinated a special issue of *Social Studies of Science* in which "multiple ontologies of water" was a key theme, stressing how "water is multiple, not only in its meanings, but more importantly, in its very materiality. . . . Its properties are not fixed. Rather, water reveals its complex, multilayered biophysical identities for particular enactments depending on assemblages that are in place or still in the making" (Barnes and Alatout, 2012 pp. 484–485). A 2017 paper by Julan Yates, Leila Harris and Nicole J. Wilson explores the political implications of different ontologies of water, highlighting "water ontologies as a site of political contest" and what they describe as "ontological disjunctures – conflicts over the very essence and being of water" (Yates, Harris and Wilson, 2017 pp. 2, 3). Sensitivity to ontological difference, they argue, "is especially needed in settler-colonial contexts where modern water ontologies remain hegemonic" (ibid., p. 3). Drawing from their research with and on First Nations peoples in Canada and particularly in British Columbia, they ask "what the implications would be to take seriously the possibility and politics of a multiplicity of water-related worlds, highlighting multiple water realities and ways of being-with-water, not just different perceptions of or knowledge systems tied to water's (singular) material existence" (ibid., p. 2).

Yates, Harris and Wilson describe how water is regarded by many elders as "lifeblood", a way of identifying water that is common among many traditional First Nations peoples in Canada and elsewhere and which differs radically from the modern water of scientists, engineers, government agencies and others who are concerned with developing, conserving and protecting "water resources." Water as "lifeblood" is fundamentally different from water as resources, connoting the sense in which water is "an animate being" that exists "in a relational connection with humans and other living beings" (ibid., pp. 4, 6). "More than a cultural construct, water-as-lifeblood has been described as a place-based, rights-producing ontology, which connects diverse beings and foregrounds water's health and vitality" (ibid., p. 6). "The Elders believe that water is alive or biotic. It has a living spirit. . . . Water still has. . . a special fundamental place in the First Nations' ecosystem—it is at its heart, since it provides the "'blood of life'" (Blackstock, 2001 p. 12).

We will return to the question of water-lifeblood as a rights-producing ontology later on. For now, let's consider this question in light of the premise of the ontological turn that reality is worlded/enacted as a process rather than pre-existing its becoming, observation or identification. In keeping with this premise, it would seem that a condition for the occurrence of water-as-lifeblood is when water is "in a relational connection with humans and other living beings" (Yates, Harris and Wilson, 2017 p. 6). I have argued that water is essentially not a thing but is rather a process that has the potential of yielding myriad outcomes depending on its relations, entanglements and engagements, including its engagements with people: "We mix language, gods, bodies and thought with water to produce the worlds and the selves we inhabit" (Linton, 2010 p. 3). Water, thus, has the potential of becoming that which is other than lifeblood. The water realized or worlded by Lavoisier in his laboratory is one example. Nor would it seem that the water NASA has identified as occurring on Mars could be considered water-as-lifeblood. In the sense we are suggesting here, water is lifeblood by virtue of its involvement in earthly life-processes, including social processes involving people. Thus, we can argue that water is brought to life by its lively relations rather than being imbued with an intrinsic or essential life-principle (see Ingold, 2007). Water-as-lifeblood is, therefore, an ontology of water that is every bit as real as H_2O, and like H_2O, it requires certain relations, practices and circumstances to be brought into being.

An example of this is the "Mother Earth Water Walk" around the North American Great Lakes. Between 2003 and 2009, Josaphine Mandamin, an Anishnaabe elder from Thunder Bay, Ontario, led numerous walkers around each of the North American Great Lakes and down the St. Lawrence River to where it meets the Atlantic Ocean. At the mouth of every stream and river tributary to the Great Lakes, Mandamin and her companions would stop and speak directly to the water, offering prayers, tobacco and thanks. Throughout their peregrinations, the walkers carried a copper bucket containing water from the lake they were circumambulating. "I've heard so many times, 'You're crazy. . .'" Mandamin was quoted as saying. "But we know it's not a crazy thing we're doing; we know it's for the betterment of the next generations" (McMahon, 2009).

This is evidently a very different kind of water from that which most of us are familiar with. One could perform a chemical analysis of the stuff in Mandamin's copper pail and it would certainly become H_2O. But for the Water Walkers, it was alive and sentient. In an interview with Canadian researchers, Mandamin confirmed: "Water can come alive if you pay attention to it. Give it respect and it can come alive" (Anderson, Clow and Haworth-Brockma, 2013 p. 16). But critically, it depends on *how* one pays attention, how one attends, because attending to – paying attention – is a world-making practice.

Paying respect to water would seem to be an important characteristic of hydrosocial relations among First Nations peoples in many parts of Canada.[4] In a recent article, Nicole Wilson and Jody Inkster write, "Rarely, have we witnessed a conversation about water or water governance in Yukon, Canada, where First Nations there have not emphasized the importance of respect for water" (Wilson and Inkster, 2018 p. 2). They emphasize that for these people, "respect" is not simply a question of "deferential regard or esteem felt or shown" but a matter of practice "characterized by reciprocal relations of responsibility between people and water as a 'more-than-human person'" (ibid., p. 2). We are arguing that water does not exist as a more-than-human person any more than it exists as a resource or even as H_2O, prior to its realization as such. Each of these renditions of water requires a form of practice, a type of relation, a kind of engagement, in order to be realized. For Lavoisier, this involved feeding water through a laboratory instrument. For Josephine Mandamin and the Water Walkers, it involved daily ritual offerings, prayers and songs, as well as the physical carrying of water over long distances around the Great Lakes.[5] And for the First Nations peoples of Yukon interviewed by Wilson and Inkster, "to 'respect' water is to engage in a manner consistent with the protocols or conventions required to maintain appropriate social relations, whether in relation to the spirit of a certain body of water or in reference to more general protocols for respecting water" (ibid., p. 11).

It follows that the accordance of rights to water itself flows from such engagements and relations. The recent granting of the legal status of *personhood* to the Whanganui River in New Zealand is cited as an example of how "decentering the role of humans in water governance involves acknowledging the rights of water itself" (Wilson and Inkster, 2018 p. 16). Indeed, headlines describing the event in 2017 proclaimed that the river had been granted the same rights as people. However, the more salient point is the ontological proposition recognizing the Whanganui River as a legal *person,* from which flow the rights accorded the river. Traditionally, a legal person is defined by having the ability to engage with others in such things as entering into contracts. In this sense, personhood flows from the river's relations and engagements with people and with myriad other living and non-living things. Arguably, rather than decentering the role of humans in water governance, according rights to the river flows from a recognition of an ontological status that is itself brought into being through the kind of social relations that the Maori have always maintained with rivers and other waterways in New Zealand (Salmond, 2014).

Political ontology

To speak and write of multiple ontologies in the way described previously is itself a political act in the sense that it strengthens the position of (the indigenous) people whose ontological claims contradict or, otherwise, oppose the modern, hegemonic ontology underpinning natural resources development. This is one objective of a project known as "political ontology" introduced by Mario Blaser, an anthropologist who currently occupies the Canada Research Chair in Aboriginal Studies at Memorial University in Newfoundland, Canada. His 2013 paper, *Ontological Conflicts and the Stories of Peoples in Spite of Europe: Toward a Conversation on Political Ontology*, built on the work of other researchers in anthropology, STS and cognate fields interested in the political implications and dimensions of the ontological turn (Blaser, 2013).[6]

As might be expected, the political dimensions of the ontological turn are described in different ways by different researchers. At the risk of overgeneralizing, for most STS scholars, politics revolves around the conundrum of how to interact in a world, or worlds, of different and competing realities. In what is perhaps its best-known formulation, the project of "cosmopolitics" proposed by Isabelle Stengers revolves around a notion of cosmos, a thoroughly indeterminate "unknown constituted by these multiple divergent worlds and... the articulations of which they could eventually be capable" (Stengers, 2005). Interpreted as "the composition of the common world" by Bruno Latour (Latour, 2014), Stenger's cosmpolitics becomes a project of common survival in times of ontological rupture, epistemological cacophony and competing fundamentalisms. "We face a situation in which, on the one hand, real peace is unattainable if negotiators leave their gods, attachments, and incompatible cosmos outside the conference room. On the other hand, a freight of gods, attachments, and unruly cosmos make it hard to get through the door into any common space" (Latour, 2004 p. 457). Less a prescription for building something in common than a statement of the challenges involved, cosmopolitics is minimally concerned with questions of power or with who gets to decide what realities get enacted and how.

The ontological politics of researchers in anthropology tend to be more concerned with the latter questions, as these researchers are often motivated by a desire to defend and legitimize the worldings of the (typically colonized and marginalized) peoples they work with. Thus, for example, Arturo Escobar champions "ontological political practice" as a form of "territorial and commons defense" (Escobar 2016, p. 19). As P. Heywood notes, most proponents of the ontological turn in anthropology "have argued that its political implications are inherently progressive: they point out that not only does the approach aim to take indigenous thought seriously in a way that cultural relativism does not, but its openness to difference also makes it fundamentally revolutionary as a way of thinking, keeping us continually on our conceptual toes" (Heywood, 2017). Blaser's "political ontology" fits into the latter as an activist form of scholarship. On his webpage, Blaser writes "The guiding insight" of his research "is that in many Aboriginal peoples' experiences and practices we can see examples of the World Social Forum's slogan, 'Another World is Possible.' In this sense, working with and learning from Aboriginal traditions opens up an avenue to address in unsuspected ways the challenges that we are facing nowadays from increasing inequality to the environmental crises" (Blaser, 2018).

A basic objective of the "political ontology" project is, therefore, to give credence to the ontological claims of indigenous peoples so as "to take others and their real difference seriously" (Blaser, 2013 p. 550). The second aspect of political ontology follows on this and offers a concrete proposal for the cosmopolitical project of finding ways of getting along constructively among and between people who espouse not just different interests or perspectives or points of view but different realities. Blaser describes making "an attempt to carve out a space to listen carefully to what other worldings propose" emphasizing that "carving out a space to listen is also carving out a space to tell another story to (and about) ourselves, to engage in other kinds of worlding that might be more conducive to a coexistence based on recognizing conflicts rather than brushing them off as irrelevant or nonexistent" (Blaser, 2013 p. 559). Thirdly, once a space has been produced for mutual listening, Blaser's approach is to find where different worldings overlap so as to enable actions that work for both, where "there might be partial co-occurrence of the entities, but the difference is not cancelled." To use Blaser's term, the aim is to enable actions that are "homonymic", addressing different things simultaneously (Blaser, 2016 p. 558). An example he gives relates to caribou in Labrador, which are one thing to Innu hunters and quite another thing to officials of the Ministry of the Environment, a situation that makes a concerted program of caribou regeneration/resource conservation impossible. Blaser in effect identifies a place where these two versions of caribou overlap in a way that would permit both Innu hunters and government officials to agree on methods that would permit the regeneration of caribou through maintaining proper human-caribou relations/conservation of the resource (Blaser, 2016).

With respect to water(s), Yates, Harris and Wilson argue along similar lines, writing "opportunities exist to overcome persistent ontological dissonance in water governance by working at the points of interconnection and overlap among these ontological starting points" (Yates, Harris and Wilson, 2017 p. 10). They develop "the notion of *ontological conjunctures*, which is based on networked dialogue among multiple water ontologies and which points to forms of water governance that begin to embrace such a dialogue" (ibid., p. 1) This "dialogue among ontologies" (ibid., p. 11), they suggest, can be fostered through "mutual co-learning," which "can be promoted through 'situated engagement with ontological pluralism,' where interactions are contextualized and concepts and practices become reliant on specific circumstances for their relevance and meaning, making it possible to acknowledge multiple epistemological and ontological perspectives" (ibid., p. 12). As an example of ontological conjuncture, they suggest that the "interconnected ontology of water-as-lifeblood" overlaps with water governance approaches that are based on watershed protection or source protection rather than end-of-pipe technologies. Thus, achieving sustainable drinking water quality would be consistent with an ontology of water as "an animate being" and a form of life itself and which, moreover, connects all living things (ibid., p. 6).

Whether as a proposal for mere dialogue or for the development of specific policies that would bridge multiple ontologies, the political ontology project begins with the requirement that all the human actors involved are prepared to entertain the legitimacy and viability of alternate or opposing worlds or realities. In the next section, we will consider whether and how the concept of rights might be invoked to help meet this requirement, using water(s) as an example.

The right to (enact) water

It might seem that the notion of multiple ontologies gives rise to political problems that cannot be resolved through standard rights-based approaches. To quote John Law, "If we live in a single. . . universe, then we might imagine a liberal way of handling the power-saturated encounters between different kinds of people [of which rights are a classic example]. But if we live, instead, in a multiple world of different enactments, if we participate in a *fractiverse*, then there will be, there can be, no overarching logic or liberal institutions to mediate between the different realities. There *is* no 'overarching.' Instead there are contingent, local and practical engagements" (Law, 2011 p. 2).

It could even be argued that "the right to water" works against multiple water ontologies. This is because declaring a right establishes a relation between the object of the right and the right-holder, with the effect of fixing terms of the relation (Linton, 2012). The right to water has often been formulated in terms of a sufficient quantity of clean water for personal, domestic needs.[7] We could argue that the idea and the actuality of things like the individual human being and modern water are sustained in part through the discourse of the right to water as it is constructed in this way. There is nothing wrong with this. Recognizing the right to water is certainly a positive development. However, it is *also* possible to use the concept of the right to water to define different kinds of relations. Elsewhere, I have suggested the possibility of formulating a right that brings people as a collective into relation with the action and capacity of water to perform certain social functions (Linton, 2012 p. 48).

Can we go further, retaining the concept of rights to help nurture multiple water ontologies? Can we imagine a right that would accord/guarantee respect for alternate ontologies of water? We might, for example, simply proclaim that everyone ought to have the right to his or her own understanding or belief of *what water is*. That this might seem ridiculous is partly because water is so thoroughly naturalized – it is taken by most to be so naturally obvious what water *is* that the notion of a right to define or understand its essence or its being *differently* seems absurd. Yet as we have seen, the identity of water has been and is varied and unstable. It has required a great deal of learning, law and institutions, including fixed rights to water, to stabilize water as H_2O and help sustain modern water as a hegemonic discourse. The current interest in multiple ontologies of water, or multiple waters, is a manifestation of the crisis of this hegemony and of growing respect for alternative versions what water is or what it could be.

One route for defending the right to uphold and assert one's belief of what water is would be to point out that there is precedent for such a right, particularly in Article 18 of the Universal Declaration of Human Rights: "Everyone has the right to freedom of thought, conscience and religion; this right includes freedom to change his religion or belief, and freedom, either alone or in community with others and in public or private, to manifest his religion or belief in teaching, practice, worship and observance." But the right to freedom of thought, granting mutual respect for different beliefs, is hardly an effective starting point for dealing with ontological pluralism. As Heywood points out, "...your interlocutor may 'believe' the tree to be a spirit, and you may 'respect' this belief as much as you wish, but your own belief is probably not what you would consider being a belief at all; it is what you would think of as 'knowledge.' You do not think of yourself as 'believing' it to be a tree, you know it to be so. 'Respect', in such a situation, becomes a synonym for mere toleration..." (Heywood, 2017).

Put another way, "belief" is a word we often use to describe a kind of difference in others that we don't understand ourselves. This suggests another problem with the notion of a right to uphold one's belief in one or another ontology and, thereby, locate the ontological difference in thought or belief; it makes ontology intrinsic to the person, or the thing itself, rather than in the relations and practices that constitute it. As we have already seen, the ontological difference is not simply a matter of belief or of one's way of thinking about the identity of something. The argument that reality is practiced, enacted or performed implies that people don't just think that water is this thing or that thing but that they are involved in the bringing forth of the thing itself. This means that if we are to retain the concept of rights to give credence to and help deal with multiple ontologies, the right has to be in respect of practices, enactments and performances. Such a right might be attached to the individual person, however, with different ontologies of water, as with most other things, the practice tends to be shared and collective. If we follow this line of thought, that water *is* lifeblood for many First Nations peoples in Canada does not mean that there is something intrinsic to these people that makes water lifeblood for them but that they share practices of "paying attention," as Josaphine Mandamin puts it, that bring water to life in particular ways. By the same token, there is nothing intrinsic to a trained chemist that makes water a compound of oxygen and hydrogen to such a person. It is by virtue of their shared material and embodied practices – including modern scientific practices – that people who constitute what we often call a particular "culture" become associated with a given ontology; the ontological difference is not inherent in the "culture" nor in the individuals who constitute it but in the practices and engagements they partake of.

Conclusion

Suggesting the right to practice, enact or perform different waters into being is hardly a prescription for a more harmonious approach to dealing with water conflicts. No amount of citizen participation, social learning or stakeholder involvement in the decision-making process is likely to resolve the problem of different waters realized through different practices. The "lifeblood-water" associated with Canada's First Nations peoples is – literally – fundamentally at odds with the water resources of the various Departments of Natural Resources or their equivalent. The project described as "political ontology" represents an effort to develop methods for doing politics among and between people who have not just different interests or perspectives or points of view but different ontologies or realities. However, except in the circumstances where there may be overlapping realities (where, for example, there are obvious "ontological conjunctures"), the right to enact alternate waters brings us down to incommensurable, perhaps irreconcilable realities, each with the potential of producing radically different material consequences. Eventually, in addition to affording possibilities for collaborative projects, the merit of acknowledging such a right might be in motivating the further intellectual and active effort required to develop a thoroughgoing politics of multiple ontologies.

More immediately, bringing multiple ontologies and rights into the same frame has the potential of bolstering the claims of First Nations and other marginalized peoples to lands and waters. A recent article by Eve Tuck and K. Wayne Yang highlights the incommensurabilities between the movement for Indigenous decolonization and various social justice movements that use the language of "decolonization" in a metaphorical sense (Tuck and Yang, 2012). Besides their radically different political projects, Tuck and Yang stress that fundamental ontological differences underlie these incommensurabilities, such as in the meaning of "land," including water, and the various relationships and interdependencies that sustain it (Tuck, McKenzie and McCoy, 2016 pp. 8–11). For Tuck and Yang, in a settler colonial context, "decolonization specifically requires the repatriation of Indigenous land and life" (ibid., p. 21). Significantly, such "repatriation" involves the assertion and acceptance of the ontological difference realized by Indigenous peoples in their engagements with the land, stressing that "simultaneous" to any such repatriation must be "the recognition of how land and relations to land have always already been differently understood and enacted; that is, *all* of the land, and not just symbolically" (ibid., p. 7). With this chapter, we are suggesting that this recognition be assured by means of a right to perform the enactments that underlie such understanding and bolster claims for repatriation. Furthermore, inasmuch as the performance of these enactments may be considered a means of achieving water security for the people involved, this would contribute to what Jepson, Wutich and Harris describe as "a right to water security" (Jepson et. al, this volume).

Finally, as noted, cultural (epistemological) relativity allows for the notion that some cultural filters or representations of the one underlying "reality" are better or more reliable – by virtue of corresponding more closely or correctly to that "reality" – than others. The "one-world world," thus, authorizes a hierarchy of knowledges with objective, scientific knowledge at the top of the list and relegating traditional and local knowledges to lesser species. Any rights that people might have to personal or collective expression, or to participation in decision-making processes, are trumped by the epistemological relativism that allows for such a hierarchy. Thus, as noted by numerous observers, traditional knowledges are seldom adequately represented, incorporated or even given a fair hearing in land claims processes and environmental and resource management practices. Considering the right to ontological difference suggests a more robust guarantee of such claims being taken seriously.

Notes

1 Cameron, de Leeuw and Desbiens (2014) discuss the ontological turn in geography in their introduction to the special issue of *Cultural Geography* on "Indigeneity and ontology".

2 A useful introduction to the ontological turn, with numerous key references, is the series titled "A Reader's Guide to the Ontological Turn" available at http://somatosphere.net/series/ontology-2
3 Modern water can be conveniently represented as H$_2$O or a compound of hydrogen and oxygen. But more importantly, it entails the assertion that all the world's myriad waters can (and should) be reduced to this entity and that this constitutes the fundamental nature of water. While a thoroughly social invention, or construction, modern water is presented as water's objective reality and, thus, discredits, or denies, the reality of alternate waters.
4 For a similar argument with respect to land, see Ingold, 2000 p. 149.
5 The importance of respectful hydrosocial practice to the Water Walk is reflected in Josephine Mandamin's journal of the 2003 walk around Lake Superior, which begins: "Each day began with a cleansing of the pail of water and the eagle staff, at 5:00 a.m., and ended approximately 6:00 or 7:00 p.m., again the cleansing with the medicines. . . . The offering of our Pipe each fourth day reminded us where we came from and connected us with our ancestors and the Creator with the tobacco offering. The water we carried in our copper pail, always reminded us of our womanly responsibilities as givers of life as Mother Earth gives us, her children, life. . . . The numerous, daily water songs that we sang are now forever embedded in nature as we saw it and were welcomed by it. . ." www.motherearthwaterwalk.com/?page_id=2174
6 Blaser draws from Annemarie Mol whose 1999 paper titled *Ontological Politics: A Word and Some Questions*, addresses some of these concerns (Mol, 1999). Isabelle Stengers "cosmopolitics" (2005) also figures in Blaser's concept of political ontology (Blaser, 2016).
7 The declaration of the United Nations Water Conference held at Mar Del Plata, Argentina, in 1977, often considered the first proclamation of the right to water, held that "all peoples. . . have the right to have access to drinking water in quantities and of a quality equal to their basic needs" (United Nations Department of Economic and Social Affairs, 1992). Since then, virtually every international statement on water supply and sanitation has invoked the human right to a (sometimes specific) quantity of clean water (see Gleick, 1998).

References

Anderson, K., Clow, B. and Haworth-Brockma, M. (2013) 'Carriers of water: aboriginal women's experiences, relationships and reflections,' *Journal of Cleaner Production*, 60, pp. 11–17.
Barnes, J. and Alatout, S. (2012) 'Water worlds: introduction to the special issue of Social Studies of Science,' *Social Studies of Science*, 42 (4), pp. 483–488.
Blackstock, M. (2001) "Water: a First Nations' spiritual and ecological perspective," *B.C. Journal of Ecosystems and Management*, 1 (1). Available at: www.siferp.org/jem/2001/vol1/no1/art7.pdf
Blaser, M. (2013) 'Ontological conflicts and the stories of peoples in spite of Europe: toward a conversation on political ontology,' *Current Anthropology*, 54 (5), pp. 547–568.
Blaser, M. (2016) 'Is another cosmopolitics possible?,' *Cultural Anthropology*, 31 (4), pp. 545–570.
Blaser, M. (2018) 'Mario Blaser.' Available at: www.mun.ca/geog/people/faculty/mblaser.php
Cameron, E., de Leeuw, S. and Desbiens, C. (2014) 'Indigeneity and ontology,' *Cultural Geographies*, 21, pp. 19–26.
Dear, P. (2006) *The intelligibility of nature: how science makes sense of the world.* Chicago, IL and London: Chicago University Press.

Escobar, A. (2016) 'Thinking-feeling with the Earth: territorial struggles and the ontological dimension of the epistemologies of the South,' *Revista de Antropología Iberoamericana*, 11 (1), pp. 11–32.

Gleick, P. (1998) 'The human right to water,' *Water Policy*, 1 (5), pp. 487–503.

Hamlin, C. (2000) 'Waters or 'Water'? Master narratives in water history and their implications for contemporary water policy,' *Water Policy*, 2, pp. 313–325.

Heywood, P. (2017) 'The ontological turn,' in Stein, F., Lazar, S., Candea, M., Diemberger, H., Robbins, J., Sanchez, A. and Stasch, R. (eds) *The Cambridge Encyclopedia of Anthropology*. Available at: www.anthroencyclopedia.com/entry/ontological-turn

Ingold, T. (2000) *The perception of the environment*. London and New York: Routledge.

Ingold, T. (2007) 'Materials against materiality,' *Archaeological Dialogues*, 14 (1), pp. 1–16.

Latour, B. (2004) 'Whose cosmos, which cosmopolitics? Comments on the peace terms of Ulrich Beck,' *Common Knowledge*, 10 (3), pp. 450–462.

Latour, B. (2014) 'Another way to compose the common world,' *HAU: Journal of Ethnographic Theory*, 4 (1), pp. 301–307.

Law, J. (2011) 'What's wrong with a one-world world.' Available at: www.heterogeneities.net/publications/Law2011WhatsWrongWithAOneWorldWorld.pdf

Linton, J. (2010) *What is water? The history of a modern abstraction*. Vancouver, Canada: UBC Press.

Linton, J. (2012) 'The human right to what? Water, rights, humans and the relation of things,' in Sultana, F. and Loftus, A. (eds) *The right to water: politics, governance and social struggles*. Abingdon: Earthscan.

Linton, J. and Budds, J. (2014) 'The hydrosocial cycle: defining and mobilizing a relational-dialectical approach to water,' *Geoforum*, 57, pp. 170–180.

McMahon, K. (2009, April 4) "A native grandmother's epic walk for the water," *The Toronto Star*. Available at: www.thestar.com/news/insight/2009/04/04/a_native_grandmothers_epic_walk_for_the_water.html

Mol, A. (1999) 'Ontological politics: a word and some questions,' *The Sociological Review*, 47, pp. 74–89.

Nadasdy, P. (2003) *Hunters and bureaucrats: power, knowledge and aboriginal-state relations in the southwest Yukon*. Vancouver, BC: UBC Press.

Nadasdy, P. (2007) 'The gift in the animal: the ontology of hunting and human-animal sociality,' *American Ethnologist*, 34 (1), pp. 25–43.

Salmond, A. (2014) 'Tears of Rangi: water, power and people in New Zealand,' *Journal of Ethnographic Theory*, 4 (3), pp. 285–309.

Sambursky, S. (1962) *The physical world of the Greeks*. New York: Collier Books.

Stengers, I. (2005) 'The cosmopolitical proposal,' in Latour, B. and Weibel, P. (eds) *Making things public: atmospheres of democracy*. Cambridge, MA: MIT Press.

Tuck, E. and Yang, K.W. (2012) 'Decolonization is not a metaphor,' *Decolonization: Indigeneity, Education and Society*, 1 (1), pp. 1–40.

Tuck, E., McKenzie, M. and McCoy, K. (2016) 'Introduction – land education: Indigenous, post-colonial and decolonizing perspectives on place and environmental education research,' in McCoy, K., Tuck, E. and McKenzie, M. (eds) *Land education: rethinking pedagogies of place from Indigenous, postcolonial and decolonizing perspectives*. London and New York: Routledge.

United Nations Department of Economic and Social Affairs (1992) *Agenda 21: the United Nations Programme of Action from Rio*. Available at: https://sustainabledevelopment.un.org/content/documents/Agenda21.pdf

Wilson, N.J. and Inkster, J. (2018) "Respecting water: Indigenous water governance, ontologies and the politics of kinship on the ground," *Environment and Planning E: Nature and Space*, 1 (4), pp. 1–23.

Yates, J. S., Harris, L.M. and Wilson, N.J. (2017) 'Multiple ontologies of water: politics, conflict and implications for governance,' *Environment and Planning D: Society and Space*, 35 (5), pp. 797–815.

6 The rights to water and food
Exploring the synergies[1]

Lyla Mehta and Daniel Langmeier

Introduction

Water is integral to human food security and nutrition, and safe water is fundamental to the nutrition, health and dignity of all (see United Nations Development Program [UNDP], 2006; High Level Panel of Experts on Food Security and Nutrition [HLPE], 2015). This, notwithstanding, the human right to water and sanitation was not explicitly acknowledged in the 1948 Universal Declaration of Human Rights (UDHR), unlike other basic rights such as the right to food (RTF), only being recognized by the UN General Assembly in 2010.[2] The emergence of the right to water was the result of a protracted struggle for many years. Until the turn of the 21st century, there still remained a lot of resistance to the human right to water on the part of some nations and corporations (see Sultana & Loftus, 2012; Mehta, 2014). Despite this long overdue global recognition, the right to water (RTW) remains conceptually ambiguous. For example, there have been heated debates about whether the RTW is compatible or not with parallel global trends of water commodification and privatization (see Sultana and Loftus, 2012). It is also still unclear what constitutes the RTW, not only in terms of the actual volume of water indicated but also on whether its narrow scope should be expanded to consider wider livelihood and survival needs beyond domestic issues.

In this chapter, we focus on the latter issue and explore a broader definition and scope for the RTW. We build on HLPE (2015) and the work of authors such as Van Koppen et al. (2017), Franco, Mehta and Veldwisch (2013) and Hellum et al. (2015) who call for an elaboration of a human rights perspective to water and food that encompasses the productive uses of water while being more interconnected than the current RTW. We argue that a broader conceptualization of the RTW is more true to how water is understood and embedded in the daily lives of local women and men around the world. Local communities rarely distinguish water for domestic and subsistence purposes. It is, thus, important for the right to water to go beyond the current domestic focus in order to embrace a more holistic definition of well-being and human survival.

By way of context, the first author of this chapter led the project team for the High-Level Panel of Experts on Food Security and Nutrition's report on water for food security and nutrition, which recommended exploring the synergies between the rights to water and food (see HLPE, 2015). In early drafts of the report, the team proposed an expansion of the right to drinking water to include productive uses and possibly ecosystem services. Such a proposal was widely criticized by key donors and also actors in the UN system, including the advisors on water and sanitation of the then Secretary-General and the former Special Rapporteur of the Human Right to Safe Drinking Water and Sanitation. The argument behind such criticisms was that many countries in the global South are currently struggling to realize the basic right to drinking water due to funding constraints: any expansion of the right – or the existence of another right that focuses on water for livelihood or productive uses – would, therefore, impede the progressive realization of the RTW and create confusion while shifting priorities and financial resources. We find this to be a weak argument. As has been documented by many scholars, while financial constraints are often mentioned as an impediment to the realization of basic rights, the main problem lies in the lack of political will, accountability and the ability of powerful actors to violate vulnerable people's basic rights with impunity (see Mehta & Ntshona, 2004; Mehta, 2014). In this chapter, we, thus, question the reluctance to adopt a more synergistic approach when both empirical research and on-the-ground realities suggest that this is necessary.

The water domain has been traditionally divided into two sectors: water supply/services and water resources management or, as the 2006 Human Development Report puts it, "water for life" and "water for production" (UNDP, 2006). Water for life refers to water for drinking and domestic purposes and is considered key for human survival. Water for production refers to water in irrigation, industry and small-scale entrepreneurial activities as well as the use of water in producing food for subsistence. This distinction, however, is highly problematic from the perspective of local users whose daily activities encompass both the domestic and productive elements of water and for whom there is little sense in separating water for drinking and washing from water for small-scale productive activities that are so crucial for survival. Empirical research demonstrates the critical role played by water for productive purposes and livelihoods, especially for poor women (Van Houweling et al., 2012; Van Koppen, 2017). Hall, Vance and Van Houweling (2014b) argue that water plays a key role in livelihood activities both in rural areas, e.g. crop irrigation or brick-making, and in peri-urban areas. Furthermore, the distinction also results in a narrow scope for the RTW, especially when compared with the right to food (RTF) (see HLPE, 2015). By drawing on practices on the ground and by turning to the capabilities and other approaches – as well as a range of legal provisions – we seek to break down such distinctions. We argue that a broader conceptualization is more true to how water is understood and embedded in the daily lives of local women and men around the world. We now turn to review the basic tenets of the rights to food and water before exploring the synergies, convergences and possible tensions between the two.

The Right to Food (RTF)

The RTF was endorsed in the Universal Declaration of Human Rights (UDHR) from 1948 and has been part of the International Covenant on Economic, Social and Cultural Rights (ICESCR) from 1976. In 2014, the UN special rapporteur on the Right to Food wrote about the "transformative potential of the Right to Food", defining it as the right of every individual, "alone or in community with others, to have physical and economic access at all times to sufficient, adequate and culturally acceptable food, that is produced and consumed sustainably, preserving access to food for future generations" (United Nations General Assembly [UNGA], 2014: 4).

The General comment No 12: The Right to Adequate Food (GC12) states that the RTF imposes on the State Parties the obligations to *respect*, to *protect* and to *fulfil*. States' obligation to "*respect* existing access to adequate food requires State parties not to take any measures that result in preventing such access" (United Nations Committee on Economic, Social and Cultural Rights [UNCESCR], 1999: para. 15). Similarly, "the obligation to *protect* requires measures by the State to ensure that enterprises or individuals do not deprive individuals of their access to adequate food" (UNCESCR, 1999: para. 15). Further, "the obligation to *fulfil (facilitate)* means the State must proactively engage in activities intended to strengthen people's access to and utilization of resources and means to ensure their livelihood, including food security" (UNCESCR, 1999: para. 15). Extrapolating on these obligations to respect, protect and fulfil the means of food production, it appears that State Parties have an obligation to protect water resources from being diverted for other purposes so that there is adequate access to water for subsistence farming and for securing livelihoods.

In the case of indigenous peoples, the realization of the RTF is dependent on recognizing not only individual rights but also upholding collective rights – the right not to be subjected to forced assimilation or destruction of their culture, their right to self-determination (by virtue of which they freely determine their political status and freely pursue their economic, social and cultural development), the rights to their lands, territories and resources, the right to maintain, control, protect and develop their cultural heritage, traditional knowledge and traditional cultural expressions and their right to non-discrimination (UNGA, 2007). Thus, in the case of communities with distinct cultural traditions – most of them small-scale producers, pastoralists, fisherfolks and so on – the call for food as a human right is intrinsically connected to the call for eliminating harmful policies and practices that prevent groups from exercising their right to self-determination (Food and Agriculture Organization of the United Nations [FAO], 2009).

Moreover, the RTF also implies that the accessibility of food must be "in ways that *are sustainable and do not interfere with the enjoyment of other human rights*" [emphasis added] (UNCESCR, 1999: para. 15). This implies that the activities and processes undertaken towards the realization of the RTF must respect the environmental limits (such as minimum flow requirements) and the carrying capacity of resources, must not be at the cost of other human rights such as the RTW (priority for drinking water and sanitation in the community) or the right to health (such as protection for agricultural workers from agrochemicals). The special recognition given in GC12 to the term sustainability when it comes to access and availability of food implies that food should be accessible for both present and future generations (UNCESCR, 1999: para. 7). This could also be relevant for water use in agriculture (see Windfuhr, 2013). For example, if water resources are overexploited leading to depletion or salinity in soils, food security cannot be sustained.

GC12 further states that "[f]inally, whenever an individual or group is unable, for reasons beyond their control, to enjoy the right to adequate food by the means at their disposal, States have the obligation to *fulfil (provide)* that right directly. This obligation also applies to persons who are victims of natural or other disasters" (UNCESCR, 1999: para. 15). To successfully implement this right, the Voluntary Guidelines on the right to food (FAO, 2005) call on states to develop strategies to realize the RTF, especially for vulnerable groups in their societies. However, as the following section will show, no such guidelines or political terms exist for the RTW.

The Right to Water (RTW)

As detailed elsewhere in this volume, on November 27, 2002, the United Nations Committee on Economic, Social and Cultural Rights adopted the General Comment No. 15: The Right to Water (GC15).[3] The Committee stressed the State's legal responsibility in fulfilling such a right and defined water as a social and cultural good, not solely an economic commodity. In July 2010, access to clean water and sanitation was recognized by the General Assembly of the United Nations as a human right. Two months later, the UN Human Rights Council affirmed by consensus that the right to water and sanitation is derived from the right to an adequate standard of living, which is contained in several international human rights treaties and that is both justiciable and enforceable (UN, 2010).

Mehta (2014) and Anand (2007) draw on Amartya Sen's capabilities approach to promote a holistic view regarding the RTW and its links with wider survival issues and livelihoods to highlight that the one cannot be guaranteed without the other (see also Jepson et al. within this volume). Capabilities refer to the "actual living that people manage to achieve" (Sen, 1999: 730). This approach focuses on "substantive freedoms – the capabilities – to choose a life one has reason to value" (Sen, 1985; 1993; 1999: 74). Thus, at the heart of this approach, one must look at the freedoms that an individual can enjoy. In his capabilities approach the focus is not on the quantity of the bundles of entitlements but instead on the principle of equality and capability to do and to be.

Sen has suggested the notion of basic capabilities which are a subset of all capabilities and encompass the freedom to do "basic" things. As Sen states, basic capabilities help in "deciding on a cut-off point for the purpose of assessing poverty and deprivation" (Sen, 1987: 109). They provide a kind of threshold, or the minimum standard required, for basic functioning and are useful for poverty/well-being analysis. When translated to water, such an approach would mean that a basic amount of water is required for basic human functioning (drinking, washing and to be free of disease), and it has, therefore, been argued that this minimum requirement for human functioning should also capture livelihood and subsistence purposes (see Mehta, 2014). This strongly resonates with work by Jepson et al. (2017 and in this volume) in which they argue that water and access to it should be understood as a hydro-social and cultural process which "include[s] the breadth and scope of water flows, water quality, and water services in all of their socio-natural complexity".

The capabilities approach refrains from outlining what exactly this minimum threshold should be as "[c]onventional, established measures and metrics do not fully reflect the unique hydrosocial conditions or historical marginalization that produce water insecurity" (Wutich et al., 2017: 7). Evidence from the water sector in setting up standards around what constitutes a "basic water requirement" highlights the variations by country and by institution. Basic water requirements have been suggested by various donor agencies, ranging from 20 to 50 litres a day, regardless of culture, climate or technology. Nevertheless, culture, climate, livelihoods, urban and rural contexts clearly do matter. The World Health Organization's (WHO) definition – again seemingly blind to context – prescribes between 20–100 litres a day (WHO, 2003) but recognises that below 50 litres can only reach a "low" level of impact and that 100 is the minimum required for basic food and personal hygiene, though this amount excludes water for productive or survival activities such as growing food (Mehta, 2014). Clearly, low-end provision takes a very narrow view of the water needs of the poor, inimical to the capability approach focus on human flourishing and freedoms, which should also take into account livelihood and subsistence needs. Also, men's and women's ability to function on the basis of the same allocation of any one resource varies dramatically. People ultimately need different basic amounts of water to enjoy the same standard in terms of capability. Take the case of South Africa, which was one of the first countries that explicitly recognized the right to water. Its Free Basic Water policy provides a minimum of 25 litres per capita per day based on a household size of eight people free to all citizens (see McDonald and Ruiters, 2005). But implementing the RTW in South Africa has been fraught with difficulties and there are huge debates regarding whether the right has had a significant impact on improving the well-being of poor South African citizens and how equitable they are (ibid., see also Flynn and Chirwa, 2005). There are further heated debates about whether the right to water is compatible or not with parallel trends of water privatization or if this not rather runs contradictory to citizens' basic right to water whilst also creating new forms of poverty and ill-being (see Flynn

& Chirwa, 2005; Loftus, 2005; McDonald & Ruiters, 2005). Last but not least, it has also led to the previously mentioned debates concerning the sufficiency of 25 litres per day per person, especially if the household number is large. All these issues further emphasise the case for an expansion of the narrow scope of the RTW and for the need to link it with the RTF.

It is important to note that like the RTF, having the RTW does not mean water will be available for free to all. Instead, it implies obligations on the part of the State to respect, protect and fulfil the RTW. Furthermore, the RTW means that market-based mechanisms, such as pricing tools, should not harm people's basic right. In water debates, the declaration of water as a human right was, therefore, a key discursive shift. For the two preceding decades, water was framed around the so-called Dublin Principles, which largely stressed water as an economic good, amongst other principles concerning participation, sustainability and gender (see Nicol, Mehta and Allouche, 2012). The long road to explicitly recognizing water as a human right has been attributed to a lack of political will and resources in this area when compared to investment in other sectors (UNDP, 2006).

> Since the poor—who suffer the most from a lack of access to improved water and sanitation services—tend to have a limited voice in political arenas, as is often argued, their claims for these services can be more easily ignored if the human right to water and sanitation is not explicit.
> (Hall, Van Koppen and Van Houweling, 2014a: 852)

Through the establishment of water as a human right, states have been obliged "as duty bearers to ensure that every citizen has affordable access to water infrastructure services for drinking, personal and other domestic uses and sanitation" (Van Koppen et al., 2017: 130).

As elaborated in GC15 on the human right to drinking water and sanitation, everyone is entitled to "sufficient, safe, acceptable, physically accessible and affordable water for personal and domestic uses" (UNCESCR, 2002: para. 2). Further, the normative content of the RTW includes "the right to maintain access to existing water supplies necessary for the right to water, the right to be free from interference, such as arbitrary disconnections or contamination of water supplies, as well as the right to a system of water supply and management that provides equality of opportunity to enjoy the right to water" (UNCESCR, 2002: para. 10). It is stressed that water should be treated as a social and cultural good and not primarily as an economic good. Also, the realization of the RTW must be done in a sustainable manner, ensuring that the right can be realized for present and future generations.

Latent potentials within General Comment 15

While the development of the human right to safe drinking water and sanitation has been largely focused on domestic water, GC15 identified other aspects of the RTW, which have remained underexplored and underdeveloped. These latent potentials include:

- The clear recognition that "water is required for a range of different purposes, besides personal and domestic uses, to realize many of the Covenant rights. For instance, water is necessary to produce food (right to adequate food) and ensure environmental hygiene (right to health). Water is essential for securing livelihoods (right to gain a living by work) and enjoying certain cultural practices (right to take part in cultural life)" (UNCESCR, 2002: para. 6).
- The inextricable linkages of the right to water to the right to the highest attainable standard of health and the rights to adequate housing and adequate food. This understanding of the right to water calls to see it in conjunction with other rights enshrined in the International Bill of Human Rights, foremost amongst them the right to life and human dignity (UNCESCR, 2002: para. 3), and the rights enshrined in the Convention on the Elimination of All Forms of Discrimination Against Women and in the Convention on the Rights of the Child (UNCESCR, 2002: para. 4).
- The development of criteria to give priority in the allocation of water resources to: i) the RTW for personal and domestic uses, ii) the RTW in connection with the RTF, iii) and the right to health to prevent starvation and disease as well as to meet the core obligations of each of the Covenant rights (UNCESCR, 2002: para. 6)
- The importance of ensuring sustainable access to water resources for agriculture to realize the right to adequate food giving particular attention "to ensuring that disadvantaged and marginalized farmers, including women farmers, have equitable access to water and water management systems, including sustainable rain harvesting and irrigation technology" (UNCESCR, 2002: para. 7).
- The importance of "taking note of the duty in article 1, paragraph 2, of the Covenant, which provides that a person may not 'be deprived of its means of subsistence', States parties should ensure that there is adequate access to water for subsistence farming and for securing the livelihoods of indigenous peoples" (UNCESCR, 2002: para. 7).
- The "Statement of Understanding accompanying the United Nations Convention on the Law of Non-Navigational Uses of Watercourses" (A/51/869 of 11 April 1997) declared that in determining vital human needs in the event of conflicts over the use of watercourses "special attention is to be paid to providing sufficient water to sustain human life, including both drinking water and water required for production of food in order to prevent starvation" (UNCESCR, 2002: footnote 8).
- The importance of protecting natural water resources from contamination by harmful substances and pathogenic microbes, including the need to taking steps on a non-discriminatory basis to prevent threats to health from unsafe and toxic water conditions (UNCESCR, 2002: para. 8).

A draft version of GC15 had a section which stated that "[t]he right to adequate food entitles an individual or group to secure the water necessary for the production of food" (quoted in Winkler, 2017: 121). However, the two resolutions on the right to water by the UN General Assembly and the UN Human Rights Commission, as well as the final GC15 by the UNCESCR, decided on a rather narrow focus on safe drinking water, personal and other domestic uses and sanitation. This represents a political prioritization, which does not adequately pay attention to other uses of water, for example, by subsistence farmers. Hall, Van Koppen and Van Houweling present three further possible explanations why the right to water for productive use has not been operationalized: i) "Water is just one input in productive activities, so benefits are more indirect", ii) "productive water uses vary and depend on highly diverse hydrological, technical, institutional, and socio-economic contexts", and iii) "water use norms and practices tend to be overlooked in developing and operationalizing human rights law" (2014a: 857–858).

Even though the Committee dropped the previously cited sentence for its final version, it still maintained a section on other uses of water such as water for subsistence farming. They directly linked this to the RTF, highlighting that efforts should be directed towards ensuring "that disadvantaged and marginalized farmers, including women farmers, have equitable access to water and water management systems, including sustainable rain harvesting and irrigation technology" (UNCESCR, 2002: para. 7 in Langford, 2005: 276). There is, thus, much scope to strengthen the interpretation and understanding of different aspects of the RTW and of its inter-linkages with the RTF.

Convergences and possible tensions

Both the general comments on the RTW and the RTF converge in the prioritization of the RTW, including sufficient water to produce food. For example, GC15 highlights how the RTW is inextricably related to the RTF and stresses that priority should be given to water resources required to prevent starvation and disease (see UNCESCR, 2002). GC12 notes the importance of ensuring sustainable access to water resources for agriculture to realize the right to adequate food. Both rights are also governed by humanitarian law and it is acknowledged that the destruction of water resources and distribution points during conflict situations can kill more people than actual weapons (see UNCESCR, 1999). International watercourse law also clearly states that in the event of conflicts, human needs must be prioritized and attention should be paid to providing sufficient water to sustain human life (water for both drinking and for producing food to prevent starvation, ibid.). Such prioritization is critical because 250 rivers on earth cross international boundaries and provide water for 40 per cent of the world's population. Both rights also suggest that State parties should ensure that there is adequate access to water for subsistence farming and for securing the livelihood needs of indigenous peoples. Furthermore, water should not be diverted for other needs at the cost of these communities.

In addition, "human rights standards stipulate that the direct and indirect costs of securing water and sanitation should not reduce any person's capacity to acquire other essential goods and services, including food, housing, health services and education" (COHRE, AAAS, SDC and UN-HABITAT, 2007: xxix). There is a tension present between such a reading and the international focus on environmental sustainability and protection of ecosystems, where the setting aside of sufficient water to maintain aquatic ecosystem functioning may be seen to be in conflict with the right to sufficient water to grow food. Should this indeed be the case, the state is obliged to ensure that the RTF is met through other means.

Often difficult choices have to be made at the individual or household level when water is scarce. For example, in rural areas, women often have to decide how much time to spend on water collection (RTW) vs. food production or fuel gathering (RTF) or whether a girl child should help collect water (RTW) or go to school (right to education). Similarly, in urban slums, poor families have to apportion the money for meeting their food requirements vs. their water needs. In such instances, rights are not in conflict with each other; rather each of these rights is being violated simultaneously.

Often at the community or society level, different rights appear to conflict with each other. At times, especially under water scarce conditions, when states attempt to fulfil the RTW obligation towards the urban community, such attempts can result in the violation of the RTF or RTW of rural communities. With increasing urban migration and related urban development, water is often diverted away from rural areas and agriculture, including subsistence production systems, to meet the urban water needs. In fact, many communities find both their RTW and the RTF violated as states pursue policies that protect the interests of more powerful stakeholders or try to fulfil the right to development of some citizens (24X7 water, energy, infrastructures) over others who need to give up their lands and water resources in this process. Finally, investments in agriculture to ensure the RTF can also affect some people's RTW or damage ecosystems.

Each of the previously described instances suggests competing trade-offs and pathways with different outcomes for different social groups. According to Windfuhr (2013), decision making needs to be determined by taking into account the priorities of vulnerable groups and their basic human needs. While domestic needs (i.e. water for drinking, bathing and hygiene) are usually given the highest priority, it is also important to prevent starvation. Accordingly, water for agriculture should first and foremost be allocated to disadvantaged and marginalized groups (something which is rarely the case in reality).

The mandate of the Special Rapporteur on the right to food also includes tracking violations. Such tracking has served as a powerful tool to counteract food related injustices, allowing the Special Rapporteur to respond to allegations with respect to violations and also enabling him or her to write to relevant governments to ask them to take action to ensure redress and accountability. The Special Rapporteur on the right to water lacks this explicit mandate and thus far has largely focused on "best practices", perhaps due to the controversial nature of the RTW and the initial resistance to its existence on the part of many powerful (corporate) players. Largely, the RTW has not been deployed to focus explicitly on water management issues or the water implications of so-called land and water grabs because of the limiting of its scope to domestic uses of water. This is in sharp contrast to the Special Rapporteur on the right to food who has, in recent years, frequently commented on land acquisitions and grabs and their impacts on local people's food security (see Franco, Mehta and Veldwisch, 2013).

It is also perhaps fair to say that food discourses tend to be more political than water ones. Take for example the concept of food sovereignty, which was developed and presented to the public in 1996 by La Vía Campesina, an international peasant organization: the Declaration on Food Sovereignty (La Vía Campesina, 1996). The idea of food sovereignty exposes wider issues of social control and power in food systems (Patel, 2009), stressing the role of the state as the guarantor of rights, as well as the sovereignty of people and their agency in managing and creating local food systems as well as their claim-making capacities (see Mann, 2017). The notion of food sovereignty also acknowledges the socially constructed nature of scarcities to land, water and food and the importance of the rights of vulnerable food producers, especially rural women (ibid.).

Bridging the rights

Some early examples that bring together the rights to water and food have emerged: these pre-date official UN discourses. Thus, in 2008, Ziganshina asserted that "the human right to water should be the basis for states' obligations to supply water not only for drinking and sanitation but also for the water-dependent livelihood needs of their residents" (Ziganshina, 2008: 117). This and similar assertions are backed up by already existing legal bodies, such as the Article 14 of CEDAW which challenges the sharp distinction between water for domestic and for productive purposes (Hellum, Kameri-Mbote and Van Koppen, 2015). As Van Koppen et al. (2017) argue, even if the current prioritization of domestic water use continues to persist, this still does not exclude the possibility for a rights-based approach regarding other uses of water. The latter also claims that GC15 "implies a right to water for livelihoods with core minimum service levels for water to homesteads that meet both domestic and small-scale productive uses" (Van Koppen et al., 2017). This has been taken up by some NGOs such as Bread for the World (Gorsboth, 2017) and, most recently, it is also reflected in the draft version of the UN Declaration on the Rights of Peasants and Other People Working in the Rural Area. Its article 21.2 states that "[p]easants and other people working in rural areas have the right to water for farming, fishing and livestock keeping and to securing other water-related livelihoods" (UN Human Rights Council, 2018: 12).

The Voluntary Guidelines on the Right to Food by the FAO notes in guideline 8 that "States should facilitate sustainable, non-discriminatory and secure access and utilization of resource" (FAO, 2005: 16) listing water as an example. The HLPE's report on water for food security and nutrition (HLPE, 2015) calls on the Human Rights Council and its Special Rapporteurs (especially those concerning water, sanitation and food) to strengthen the realization of the right to water and to explore the implications of the linkages between water and food security and nutrition on the realization of human rights.

While discussing water as a human right in the Middle East and North Africa (MENA), Brooks (2007) argues that this concept is particularly old in MENA and goes back at least to the Code of Hammurabi in Babylon. The UN focus on household and domestic uses of water are, therefore, unhelpful in a region characterized by sharp competition where average annual supply for the region is well under 1500 cubic metres per capita. Thus, he advocates consideration of the RTW to grow food and the RTW for a liveable environment (that is, water to support the ecosystem) (ibid.). Water's greatest use, Brooks argues, is for growing food and everybody should have a right to a sufficient quantity of water of decent quality "to enable the growing of enough nutritious food for a healthy life" (Brooks, 2007: 233).

Some countries have already demonstrated the ability to bridge such concerns. Thus, Bolivia combines the right to water and food in one article in its 2009 constitution: "Article 16.I. Everyone has a right to water and food".[4] Ecuador, furthermore, states in its constitution that energy production cannot be promoted in opposition to the right to food and water.[5] In a comment on the right to water in Ecuador, the Ombudsman Office writes: "The human right to water, due to its transversal nature in regards to other rights, especially the right to an adequate life, is directly linked to food, as human beings need water and food to survive and to guarantee adequate health. In this sense, there is a direct relationship between land tenure and access to water – understood as factors of production – since both are indispensable to a right to food".[6]

Hellum, Kameri-Mbote and Van Koppen (2015) list further countries that have joined the rights in their constitution. "By grouping of the right to water together with other social and economic rights (most importantly the right to food and health), the ICESR, the Maputo Protocol, and the Kenyan, South African, and Zimbabwean constitutions imply a right to affordable and available water for personal, domestic and livelihood uses. All these international, regional, and national instruments recognize the indivisibility of human dignity, social justice, equality, and non-discrimination and protection of the poor and marginalized as basic principles" (Hellum, Kameri-Mbote and Van Koppen, 2015: 18–19). In Switzerland too, the right to water is considered a precondition to the enjoyment of the right to food (SDC, 2008).[7]

Despite this constitutional recognition, in reality, tensions between water for agriculture, urban use, mining and industry abound and often get in the way of ensuring the water and food security of indigenous and poorer populations (for Ecuador see Mena-Vásconez, Vincent and Boelens, 2017). Nevertheless, a human rights perspective to water, land and food can allow for legal claims and local struggles, which could slowly help with the realization of the right to water for livelihoods/subsistence (see Clark, 2017).

Conclusions

This chapter has focussed on exploring synergies between the rights to water and food. A core argument has been that a broader conceptualization of the right to water is more true to how water is understood and embedded in the daily lives of local women and men around the world. Local communities rarely distinguish water for domestic and subsistence purposes. It is, thus, important for the right to water to go beyond its current domestic focus in order to embrace a more holistic definition of well-being and human survival. It is also important for states to focus on the more productive uses of water and how these are key to the well-being and survival of vulnerable communities, protecting them against dispossession arising through land and water grabs. We have also demonstrated how some countries and legal provisions are already embracing this broader scope even while acknowledging tensions and competing trade-offs around broadening the scope. Finally, the chapter has shown how food debates have been broader, and more political and inclusive, than water ones. The latter have often been rather apolitical, not pushing for wider interpretations. This distinction also characterises the work of the Special Rapporteurs for water, tending to promote a more narrow focus within the RTW.

As stated in the Introduction, attempts to link the two rights (including on the part of one of the authors) have so far not been very successful. However, we recall the initial resistance to the right to water in the 1990s and around the turn of the century, which was then reversed when General Comment 15 emerged explicitly, providing an authoritative interpretation of the right to water. The ability to overcome such resistance shows that ideas often rejected as utopian and impractical are realized when the time is ripe. Whether the right to water can be expanded to look at subsistence and productive uses or whether there is a need for a separate human right for water for livelihood/subsistence purposes still needs to be worked out. However, we end this article in the hope that the key global players such as the World Committee on Food Security, the Human Rights Council and also the Special Rapporteurs on water, food and health will seriously explore pushing the linkages and synergies between the rights to water and food in order to ensure healthy and productive lives for all.

Notes

1 We thank Alex Loftus and Farhana Sultana for their useful comments and patience. This article builds on Lyla Mehta's work for the High Level Panel of Experts on Food Security and Nutrition's report on water for food security and nutrition (HLPE, 2015). She thanks the Steering Committee and Secretariat of the HLPE and in particular, Theib Oweis, Claudia Ringler, Barbara Schreiner and Shiney Varghese, her co-authors of the project team.

2 The right to water and to sanitation were jointly recognized by the 2010 UN General Assembly. In effect they are two different rights following the position of the Special Rapporteur on the Human Right to Safe Drinking Water and Sanitation (de Albuquerque, 2012: 27). In this chapter, we largely focus on the human right to water and not on sanitation issues. As a shorthand we use the Right to Water (RTW).
3 General Comment 15, The right to water (arts. 11 and 12 of the International Covenant on Economic, Social and Cultural Rights), U.N. Doc. E/C.12/2002/11 (Twenty-ninth session, 2002).
4 Translated by Daniel Langmeier, www.mindef.gob.bo/mindef/sites/default/files/nueva_cpe_abi.pdf
5 Art. 15, www.oas.org/juridico/pdfs/mesicic4_ecu_const.pdf
6 Translated by Daniel Langmeier, http://repositorio.dpe.gob.ec/bitstream/39000/119/1/IT-005-EL%20AGUA%20COMO%20UN%20DERECHO%20HUMANO.pdf Page 39
7 www.eda.admin.ch/dam/deza/en/documents/themen/staats-wirtschaftsreformen/170500-human-rights-approach-water-sanitation_EN.pdf

References

Anand, P. (2007) 'Right to water and access to water: an assessment', *Journal of International Development*, 19, pp. 511–526.

Brooks, D. (2007) 'Human rights to water in North Africa and the Middle East: what is new and what is not; what is important and what is not', *International Journal of Water Resources Development*, 23 (2), pp. 227–241.

Clark, C. (2017) 'Of what use is a deradicalized human right to water?', *Human Rights Law Review*, 17 (2), pp. 231–260.

Centre on Housing Rights and Evictions (COHRE), American Association for the Advancement of Science (AAAS), Swiss Agency for Development and Cooperation (SDC) and United Nations Human Settlements Programme (UN-HABITAT) (2007) *Manual on the right to water and sanitation*. Available at: www.worldwatercouncil.org/fileadmin/wwc/Programs/Right_to_Water/Pdf_doct/RTWP__20Manual_RTWS_Final.pdf

de Albuquerque, C. (2012) *On the right track: good practices in realising the rights to water and sanitation*. Lisbon, Portugal. Available at: www.worldwatercouncil.org/sites/default/files/Thematics/On_The_Right_Track_Book.pdf

Flynn, S. and Chirwa, D.M. (2005) 'The constitutional implications of commercialising water in South Africa', in McDonald, D. and Ruiters, G. (eds) *The age of commodity: water privatization in Southern Africa*. London: Earthscan.

Food and Agriculture Organization of the United Nations (2005) *Voluntary guidelines to support the progressive realization of the right to adequate food in the context of national food security* (E/C.12/1999/5). Rome, Italy. Available at: www.fao.org/3/y7937e/y7937e00.pdf

Food and Agriculture Organization of the United Nations (2009) *The right to adequate food and indigenous peoples: How can the right to food benefit indigenous peoples?* Rome, Italy. Available at: www.fao.org/3/a-ap552e.pdf

Franco, J.C., Mehta, L. and Veldwisch, G.J. (2013) 'The global politics of water grabbing', *Third World Quarterly*, 34 (9), pp. 1651–1675.

Gorsboth, M. (2017) *Wasserreport: Die Welt im Wasserstress – Wie Wasserknappheit die Ernährungssicherheit bedroht (Water report: the world suffers from water stress – how water scarcity threatens food security)*. Berlin, Germany. Available at: www.brot-fuer-die-welt.de/fileadmin/mediapool/2_Downloads/Fachinformationen/Analyse/Analyse_49_Wasserreport.pdf

Hall, R.P., Van Koppen, B. and Van Houweling, E. (2014a) 'The human right to water: the importance of domestic and productive water rights', *Science and Engineering Ethics*, 20 (4), pp. 849–868.

Hall, R.P., Vance, E. and Van Houweling, E. (2014b) 'The productive use of rural piped water in Senegal', *Water Alternatives*, 7 (3), pp. 480–498.

Hellum, A., Kameri-Mbote, P. and Van Koppen, B. (eds) (2015) *Water is life: women's human rights in national and local water governance in southern and eastern Africa*. Harare, Zimbabwe: Weaver Press.

High Level Panel of Experts on Food Security and Nutrition (2015) *Water for food security and nutrition: a report by the High Level Panel of Experts on Food Security and Nutrition of the Committee on World Food Security*. Rome, Italy. Available at: www.fao.org/fileadmin/user_upload/hlpe/hlpe_documents/HLPE_Reports/HLPE-Report-9_EN.pdf

Jepson, W., Budds, J., Eichelberger, L., Harris, L., Norman, E., O'Reilly, K., Pearson, A., Shah, S., Shinn, J., Staddon, C., Stoler, J., Wutich, A. and Young, S. (2017) 'Advancing human capabilities for water security: a relational approach', *Water Security*, 1, pp. 46–52.

Langford, M. (2005) 'The United Nations concept of water as a human right: a new paradigm for old problems?', *International Journal of Water Resources Development*, 21 (2), pp. 273–282.

La Vía Campesina (1996) 'The right to produce and access to land – food sovereignty: a future without hunger'. Available at: www.acordinternational.org/silo/files/decfoodsov1996.pdf

Loftus, A. (2005) 'Free water as a commodity: the paradoxes of Durban's water service transformation', in McDonald, D. and Ruiters, G. (eds) *The age of commodity: water privatization in Southern Africa*. London: Earthscan.

Mann, A., (2017) 'Food sovereignty and the politics of food scarcity', in Dawson, M.C., Rosin, C. and Wald, N. (eds) *Global resource scarcity: catalyst for conflict or cooperation?* Oxon and New York: Routledge.

McDonald, D. and Ruiters, G. (eds) (2005) *The age of commodity: water privatization in Southern Africa*. London: Earthscan.

Mehta, L. (2014) 'Water and human development', *World Development*, 59, pp. 59–69.

Mehta, L. and Ntshona, Z. (2004) *Dancing to two tunes? Rights and market-based approaches in South Africa's water domain* (Sustainable Livelihoods in Southern Africa Research Paper Series, no. 17). Available at: www.ircwash.org/sites/default/files/Mehta-2004-Dancing.pdf

Mena-Vásconez, P., Vincent, L., Vos, J. and Boelens, R. (2017) 'Fighting over water values: diverse framings of flower and food production with communal irrigation in the Ecuadorian Andes', *Water International*, 42 (4), pp. 443–461.

Nicol, A., Mehta, L. and Allouche, J. (2012) 'Introduction: 'some for all rather than more for some'? Contested pathways and politics since the 1990 New Delhi statement', *Institute of Development Studies Bulletin*, 43 (2), pp. 1–9.

Patel, R. (2009) 'Food sovereignty', *Journal of Peasant Studies*, 36 (3), pp. 663–706.

Swiss Agency for Development and Cooperation (2008) *A human rights-based approach to water and sanitation*. Available at: www.eda.admin.ch/dam/deza/en/documents/themen/staats-wirtschaftsreformen/170500-human-rights-approach-water-sanitation_EN.pdf

Sen, A. (1987) *On ethics and economics*. Oxford: Basil Blackwell.

Sen, A. (1999) *Development as freedom*. Oxford: Oxford University Press.

Sultana, F. and Loftus, A. (2012) *The right to water: politics, governance and social struggles* (Earthscan Water Text Series). London and New York: Routledge.

United Nations (2010) 'General Assembly adopts resolution recognizing access to clean water, sanitation as human right, by recorded vote of 122 in favour, none against, 41 abstentions'. Available at: www.un.org/press/en/2010/ga10967.doc.htm

United Nations Committee on Economic, Social and Cultural Rights (1999) *CESCR General Comment No. 12: The Right to Adequate Food (Art. 11), E/C.12/1999/5*. Available at: www.fao.org/fileadmin/templates/righttofood/documents/RTF_publications/EN/General_Comment_12_EN.pdf

United Nations Committee on Economic, Social and Cultural Rights (2002) *General Comment No. 15: The Right to Water (Arts. 11 and 12 of the Covenant), E/C.12/2002/11*. Available at: www.globalhealthrights.org/wp-content/uploads/2013/10/CESCR-General-Comment-No.-15-The-Right-to-Water.pdf

United Nations Development Programme (2006) *Beyond scarcity: power, poverty and the global water crisis*. Basingstoke: Palgrave.

United Nations General Assembly (2007) *Resolution adopted by the General Assembly on 13 September 2007–61/295. United Nations Declaration on the Rights of Indigenous Peoples, A/RES/61/295*. Available at: https://undocs.org/A/RES/61/295

United Nations General Assembly (2014) *Report of the Special Rapporteur on the right to food, Olivier De Schutter – final report: the transformative potential of the right to food, A/HRC/25/57*. Available at: https://documents-dds-ny.un.org/doc/UNDOC/GEN/G14/105/37/PDF/G1410537.pdf?OpenElement

United Nations Human Rights Council (2018) *Draft United Nations declaration on the rights of peasants and other people working in rural areas, A/HRC/WG.15/5/3*. Available at: https://undocs.org/A/HRC/WG.15/5/3

Van Houweling, E., Hall, R., Diop, A.S., Davis, J. and Seiss, M. (2012) 'The role of productive water use in women's livelihoods: evidence from rural Senegal', *Water Alternatives*, 5 (3), pp. 658–677.

Van Koppen, B. (2017) *Rural women's rights to water for health, food and income*. Rome, Italy. Available at: www.ngocsw.org/wp-content/uploads/2017/10/Rural-women_s-rights-to-water-for-health-food-and-income.pdf

Van Koppen, B., Hellum, A., Mehta, L., Derman, B. and Schreiner, B. (2017) "Rights-based freshwater governance for the twenty-first century: beyond an exclusionary focus on domestic water uses", in Karar, E. (ed) *Freshwater governance for the 21st Century*. Cham, Switzerland: Springer, pp. 129–143. Windfuhr, M. (2013) *Water for food: a human rights obligation*. Berlin: German Institute for Human Rights.

Windfuhr, M. (2013) *Water for food: a human rights obligation*. Berlin, Germany: German Institute for Human Rights. Available at: www.institut-fuer-menschenrechte.de/uploads/tx_commerce/Study_Water_for_Food_a_Human_Rights_Obligation.pdf

Winkler, I.T. (2017) "Water for food: a human rights perspective", in Langford, M. and Russell, A.F.S. (eds) *The human Right to water: theory, practice and prospects*. Cambridge: Cambridge University Press.

World Health Organization (2003) *The right to water* (Health and Human Rights Publication Series, no. 3). Geneva, Switzerland: World Health Organization.

Wutich, A., Budds, J., Eichelberger, L., Geere, J., Harris, L.M., Horney, J.A., Jepson, W., Norman, E., O'Reilly, K., Pearson, A.L., Shah, S.H., Shinn, J., Simpson, K.,

Staddon, C., Stoler, J., Teodoro, M.P. and Young, S.L. (2017) 'Advancing methods for research on household water insecurity: studying entitlements and capabilities, socio-cultural dynamics, and political processes, institutions and governance', *Water Security*, 2, pp. 1–10.

Ziganshina, D. (2008) 'Rethinking the concept of the human right to water', *Santa Clara Journal of International Law*, 6, pp. 113–128.

7 Water-security capabilities and the human right to water

Wendy Jepson, Amber Wutich and Leila M. Harris

Introduction

The human right to water discourse offers a powerful mooring for social and political claims by policymakers, public health advocates, development campaigners, and communities struggling for equitable, affordable, and safe water provision (and sanitation). International legal architecture and obligations, including the 2010 United Nations (UN) Resolution (A/RES/64/292 28 July 2010), paired with a constitutionally recognized right to water in a number of country contexts (e.g., South Africa, Bolivia, Uruguay, Ethiopia), reflects a growing global consensus and resonance of its core claim—the idea of universal access to safe and affordable water regardless of ability to pay. For many, the recognition by the UN General Assembly in 2010 signified a clear victory for water justice advocates. The human right to water policies and discourses that strive to attain safe and affordable water have influenced policymakers, shaped international development targets, and appealed to millions around the world. As Pope Francis (2017) stated during his *Dialogue on the Human Right to Water*,

> The right to water is essential for the survival of persons and decisive for the future of humanity. . . Respect for water is a condition for the exercise of the other human rights. . . Our commitment to give water its proper place calls for developing a culture of care (cf. ibid., 231) and encounter. . .

The call for and meaningful work towards conceptualizing and achieving a human right to water reveals its salience for imagining, debating, and broadening the policy space for alternative concepts and visions of water governance, particularly those that embrace equity, justice, and sustainability. Whether underscoring the relationship between to the human right to water or other ongoing water governance shifts, it is clear that we will be debating the meaning of the right to water and *how* to achieve these goals, not if we should, for years to come.

Notwithstanding the aspirational dimension of the human right to water discourse and its policy success, many scholars have identified key shortcomings, many of which have been well-reviewed in the literature. The human right to water has been criticized as individualistic, state-centered, anthropocentric, and vulnerable to co-optation by the private sector (Bakker, 2007; Bustamante, Crespo, & Walnycki, 2012). Some have also argued that an emphasis on "drinking water" as part of basic-needs orientations is problematic given fundamental differences cross-culturally and for various livelihood strategies (Goff and Crow, 2014). Linked to this, the goal of water access and quality alone pays little attention to broader associated considerations such as human dignity (Redfield & Robins, 2016; Morales, Harris, & Öberg, 2014). A narrow focus on basic needs also risks overlooking water insecurities in the Global North (Ranganathan & Balazs, 2015; Jepson & Vandewalle, 2016). Urban struggles for water in Flint or Detroit in the United States, for example, are often portrayed in the as outcomes of technical failures (Morckel, 2017) rather than as political failures to fulfill human rights obligations (Clark, this volume).

Yet, the sharpest critiques do not negate clear benefits and salience of human right to water calls as a discursive repertoire and policy frame. Rather than reject the notion as co-opted or post-political, scholars have called for new pathways to reformulate and re-politicize the human right to water (Sultana & Loftus, 2012; 2015; Perera, 2015). Others call for a holistic conceptualization of the human right to water so as to fulfill its potential by focusing on lived experiences, unevenness, and water's materiality (Rodina, 2016).

Angel and Loftus (2017) interrogate the paradoxical role of the state within the human right to water debates and policy implementation. They argue that the UN recognition of the human right to water and its incorporation into legal frameworks has re-centered the state in these struggles, making ground gained by activist communities vulnerable to co-optation, dilution, and denial by state power. According to these authors, a new vision of the human right to water requires a fundamental rethinking of the state—one that proactively adopts a relational approach to the state as co-produced through socio-natural processes. They argue that the human right to water similarly needs to be relational and that scholars, activities, and communities need to "think and act within-against-and-beyond the right to water" to advance political change within (and without) state processes while also holding in tension that such water politics can yield new possibilities for progressive socio-ecologies (Angel and Loftus, 2017, p. 7).

This chapter seeks to build on these calls to think through the possibilities within-against-beyond the human right to water in tandem with re-centering lived experiences of water. Notably, we draw inspiration from the capabilities approach to incorporate a broader understanding of hydro-social relations. This is a clear departure from previous utilitarian approaches to water security that prioritize water access and environmental and economic services (Schmidt, 2012; Zeitoun et al., 2016). It is a distinct departure from Grey and Sadoff's definition of water security (2007, 547–48), which attends to water availability, risk (Garrick and Hall, 2013), or water security discourses framed in terms of geopolitics, economic development, large infrastructure, or state policy and governance (Zeitoun, 2011; Lankford, Bakker, Zeitoun, & Conway, 2013; Scott et al., 2013). Our definition of water security is grounded in the lived experience.

Attention to the community, household, and individual experiences of hydro-social relations provides clarity as to the societal consequences that circulate through the dynamic socioeconomic change in water flows and its meaning in relation to everyday life and human well-being. This chapter presents concepts and analytics associated with what we term a "water-security capabilities approach." We propose that water security needs to be redefined, and most importantly, that a redefinition should shift from securing water as a thing and end in itself (as H_2O) to securing water relations that recognize the wider relations through which water shapes people's lives and contributes to human flourishing and well-being (Jepson et al., 2017). We review the capabilities approach and water security to situate our analysis, followed by our response to two questions: (1) How does a water-security capabilities approach advance current conceptualizations of the human right to water? (2) What are the implications of a water-security capabilities approach for water equity?

The emergence of a water-capabilities approach

The capabilities approach (CA) is a conceptual framework that develops core concepts of well-being, freedom to achieve well-being, development, and social justice. Originating in welfare economics and political philosophy, CA focuses on how social arrangements contribute to, or detract from, human flourishing and freedom (Nussbaum & Sen, 1993; Nussbaum, 1997; 2011b; Sen, 2000). CA purports that freedom to achieve well-being is a matter of what kind of life one is effectively able to lead and the pathways and mechanisms through which an individual (or a community) is able to define and achieve well-being. Early capabilities approaches have been subject to criticism of individualism; however, recent work with indigenous communities has demonstrated the ways CA enables community functioning (Schlosberg and Carruthers, 2010).

Several core foci cut across CA interdisciplinary scholarship. First, capabilities are understood as what people are able to do and be, or the genuine (and positive) freedoms and opportunities to realize functionings. Capabilities reflect the fact of whether a person could do something if he or she wants to; that is, capabilities are considered *intrinsic* (not instrumental) aspects of well-being that a person may value and are constituent of human dignity. Functioning is defined by what a person actually does or is, such as nourishment, taking part in a religious community or engaging in political life. That is, functionings are the corresponding achievements to capabilities and are constituent to human existence.

While CA's intellectual roots can be traced to Amartya Sen, who first reframed human development in terms of well-being, and Martha Nussbaum, who advances a theory of social justice (Nussbaum, 2003; 2011b), the concept resonates across development studies, welfare economics, gender studies, and sociology. Capabilities analyses address a wide range of topics including the advancement of social justice, policy, quality of life measurements, and social arrangements and institutions. CA respects people's different ideas of the good life and their capacity to achieve it. Moreover, human diversity is stressed through the recognition of unique profiles of conversion factors. Individuals and communities have different abilities to convert resources—from goods and services to educational attainment—into functionings. These *conversion factors*, which have been categorized as personal, sociocultural, and environmental, provide a fuller picture of available resources—owned or accessible—to realize well-being and achieve functionings (Robeyns, 2018). Notably, as Nussbaum argues, "the language of capabilities enables us to bypass the troublesome debate"—about the European origins of rights—because "when we speak about what people are able to do and to be, we do not give the appearance of privileging a Western idea" (Nussbaum, 2003). CA also allows for value pluralism in human development, eschewing the tendency to reduce development to one metric, such as wealth. Nussbaum argues that the openness of CA offers an alternative narrative to the "poverty-prosperity" paradigm or utilitarianism. Certainly, the means of well-being are important—such as resources, housing, social institutions, etc.—but attention to capabilities underscores that these are *not the ends of well-being*. As we explore here, these elements of CA all have potential importance for reformulating the human right to water debate.

Capabilitarian scholars have recently turned their attention to the analysis of ecological and environmental issues, including ideas of natural capital and ecological limits, environmental justice, and animal ethics (Nussbaum, 2011a; Rauschmayer, Omann, & Frühmann, 2012; Pelenc & Ballet, 2015; Ballet, Marchand, Pelenc, & Vos, 2018). Particularly relevant for our interests, recent examples have also addressed access to water. For example, Goldin (2013, p. 315) argues that a capabilities approach to the water sector encompasses various dimensions, including human health and goods, education and literacy, significant relations with others, participation in social life, self-determination and autonomy, accomplishment, aspiration and self-respect, and basic mental and physical functionings. In this way, water is understood as a conversion factor, a resource necessary to achieve wellbeing (cf., Anand, 2007). Along similar lines, Lyla Mehta (2014) argues that the right to water is necessary for human capabilities—whether in service to support health, bodily requirements as well as productive resources for different livelihoods; thus, she advances an ethical claim on the state for the provision of water for all.

Jepson et al. (2017) advance a capabilities framework to re-conceptualize the contested concept of "water security" for human development. Water security understood through a CA framework necessarily attends to water as part of a hydro-social process that is relational, based on negotiation and interaction at individual and collective scales. A central capability is the ability to engage with and benefit from sustained hydro-social processes that include the breadth and scope of water flows, water quality, and water services in all of its socio-natural complexity. Similar themes are taken up by Gimelli, Bos, and Rogers (2018), who are interested in examining current international metrics of water access in terms of the capabilities approach. Most importantly, applying a CA framework to water security issues incorporates a relational framework to analyze broader considerations and pathways important for hydro-social relations. For our purposes, we argue that water is not simply a resource (or "conversion factor" in CA terms) but should be understood and analyzed as *co-produced hydro-social flows, services, and meanings* that support the achievement of human well-being *as defined by individuals and communities*. Thus, the water-security capabilities approach offers new pathways to re-conceptualize the human right to water.

Water-security capabilities approach and revisioning the human right to water

A long-standing dialogue between capabilitarian perspectives and human rights scholarship addresses themes of duty, obligation, and the implementation of human rights into policy and practice (Nussbaum, 1997, 2003; Sen, 2005). There are salient aspects to this dialogue that inform our analysis of the human right to water from a water-security capabilities perspective. First, there is a recognition that the language of rights plays an important political role in the attainment of justice. Second, capabilities—whether on Nussbaum's endorsed capabilities list or implied in terms of Sen's "process-freedoms"—meet the threshold of urgent importance similar to human rights: "that just by virtue of being human, a person has a justified claim to have the capability secured to her," similar to a human right (Nussbaum, 1997, p. 293). In this way, capabilities are seen as fundamental entitlements, *equivalent* to "first-generation rights" (civil and political liberty) and "second-generation rights" (economic and social). Finally, freedom and choice ground capabilitarian views of human rights, therefore the ideas of what is a capability (and in this view a human right) are flexible, and open to revision and rethinking, which, by default, demands public deliberation to enable progressive realization of these dynamic goals (Nussbaum, 2011b).

It is against this backdrop that we focus on how water-security capabilities approach can advance current debates on the human right to water in terms of two central questions: (a) *to what* do people or communities have a right and (b) *whose* duty or obligation is it to ensure the right or capability? Our response to each question, then, allows us to "think and act within-against-and-beyond the right to water" with the benefit of our broadened orientation not on water as an object but a more relational and processual focus on hydro-social relations and human well-being.

First, scholars and policymakers need to address the fundamental question: *to what do people or communities have a right*? The human right to water pivots on what Linton (2010) calls "modern" water. A narrowly defined understanding of the human right to water, which entitles everyone to sufficient, safe, acceptable, physically accessible, and affordable water (H_2O) for personal and domestic uses is silent to the complexity of human-environment interactions. One can argue that ecosystem services are of central importance to the realization of the human right to water. For example, the UN acknowledges that the loss of key ecosystem services, such as water and biodiversity, undermine human rights, for example by reducing agricultural and fisheries outputs, negatively affecting health or removing natural filters in the water cycle (United Nations High Commissioner for Refugees [UNHCR], 2018). But in order for this to be a universal positive right, the HRW frames water as a resource and individuals as consumers rather than agents in complex socio-natures.

The water-security capabilities perspective opens new visions of what a right to water is. First, following from earlier discussions in Jepson et al. (2017), the claim we are making is that a right to water needs to be redefined as "a right to water security," defined as the ability of individuals, households, and communities to benefit from sustained hydro-social and cultural processes that include the breadth and scope of water flows, water quality, and water services in all of their socio-natural complexity. This perspective respects the freedom to engage and benefit from hydro-social and cultural relationships. Water-security capabilities are considered *intrinsic* not instrumental aspects of well-being that a person or community may value. That is, we propose that water security as a capability, not a conversion factor.

Moving forward, a capability approach to the human right to water also opens space to integrate multiple water ontologies into its frame (Shah, Angeles, & Harris, 2017; Linton, this volume), taking seriously "the possibility and politics of a multiplicity of water-related worlds" (Yate, Harris, & Wilson, 2017). The right to water security from a capabilities perspective provides political and philosophical grounds for individuals and communities to assert certain realities of what water is and how they self-define and understand water in terms of their fluid socio-environmental relations with water systems, flows, and services. Moreover, the water-security capabilities approach to recognizes *co-constituent well-being* between society and water flows. That is, a water-security capability approach acknowledges complex socio-natures as a constituent to human capabilities and, thus, it offers a unified theoretical entry point and ethical claim, not a separate eco-centric claim, for ecological processes to be fundamental for the human right to water. Thus, a water-capabilities approach opens a conceptual and *political* pathway for diverse water ontologies to bear on claims and debates over the human right to water.

The second question relates to *obligation and duty*, in particular, the role of the state in securing the human right to water. The role of the state and the expectation of government involvement in achieving the prescriptive claim underscores a primary difference between the two approaches. First, the human right to water is considered by some to be state-centric (Bakker, 2007; Angel & Loftus, 2017). The human rights claim pivots on legal pathways for the redress of marginalized and impoverished communities to make demands on the state for improved water access or conditions. As a designated human right, it creates corresponding legal obligations to ensure the enjoyment of the right, and that obligation lies with each state to follow the necessary technical, economic, and infrastructural steps to work towards the progressive realization of the right for its population. While many are critical of the state-centricity of the human right to water, given common failures of the state, some have noted nonetheless that the state-centricity also takes on particular meaning, and potential importance, given movements away from the state as part of ongoing market pressures and neoliberal restricting of water governance of the past several decades (Mirosa and Harris, 2012).

From a capabilities approach to water security, we reconsider the implications in terms of the role and meanings of the state. To the extent that the capability approach respects peoples' different ideas of the good life and given that capabilities are the goals (as a relation and process) rather than focusing on any particular single outcome, this has the potential to move the discussion away from state responsibility to secure water as an object. Instead, the approach foregrounds processes through which states, communities, and individuals define water capabilities and, in turn, how these definitions serve as a basis to make claims on the state. With the broadened focus on hydro-social relations, communities and individuals could potentially press for a variety of claims related to healthy, equitable, and sustainable hydro-social relations—including cultural and political recognition, political participation, collective action, and democratic rights. In this recasting, the role of the state is not *necessarily* (or only) to provide H_2O but rather to facilitate or help citizens realize the right to participate and engage in social and political collective action and maintain and secure sustainable and equitable hydro-social relations in all of its complexity. Thus, a water-security capabilities approach reinforces individuals and communities as citizens and political actors rather than reduced to only consumers of water.

Recasting state-society relations through a renewed approach to the human right to water is crucial. As we see in the case of Katie Meehan's work in Tijuana, Mexico, residents' use of alternative rainwater collection systems offers opportunities for autonomy *from* the state (Meehan, 2014). In Jepson's ongoing research in Fortaleza, Brazil, we have also documented systemic self-disconnection from state-owned water utilities for individual and even collective groundwater wells, a source viewed by urban residents as more reliable and secure. Moreover, state-imposed systems—in the pursuit of the human right to water—have been shown in some cases to undermine existing water sharing regimes. With an open-ended appreciation of water-security capabilities, in lieu of a narrow framing of the human right to water as securing rights to H_2O, we can enable an appreciation of the complex and often deeply ingrained in senses of reciprocity between family, community, and non-human natures, all of which can be critical for how we understand the senses of well-being and human flourishing (Wutich et al., 2018). Therefore, a narrow state-driven and state-centric realization of the human right to water without attending to cultural obligations, expectations, and fundamental social relations could unduly impinge or constrain freedoms to realize what a person or community is, or does, in relation to water flows and systems. Duty and obligation to realize or ensure the human right to water (security) is coproduced and tied to inclusive forms of water governance that are within-against-and-beyond the state.

Engaging CA more meaningfully in these discussions helps to amplify these emergent interactions, and it also illuminates some of the specific pathways that might be followed to reveal and enliven diverse pathways to water-security capabilities, as differentiated in various communities, across time and space.

Equity and water-security capabilities approach

Yet if a capabilities approach provides for water security to be achieved through diverse and contingent pathways, what are the implications of a water-security capabilities approach for equity? Equity lays at the heart of conflicts over water, and some have also argued that conflict is a necessary precondition for both sustainable transition and justice in water systems (Ingram et al., 2008, p. 8). In riverine systems, for example, conflicts between upstream and downstream users are ubiquitous and a major focus of institutional rule-making. In addition to concern with how people and places are unequally affected by water-related changes, quality, and access concerns, it has also been suggested that equity plays an important role in scholarly framings of water problems, water justice struggles, and water governance deliberations (Perreault, 2014; Zwarteveen & Boelens, 2014; Boelens, Perreault, & Vos, 2018; Sultana, 2018).

While water scholars agree that equity matters, there is relatively little consensus on what equity means, how to recognize it when we see it, or how to most ably promote it in water governance frameworks (Lu, Ocampo-Raeder, and Crow, 2014). Rather, scholars emphasize the historical and cultural contingency of understandings of equity (Lauderdale, 1998; Ingram, Whiteley, & Perry, 2008). Even cross-cultural analyses yield little overarching agreement as to how equitable water distribution is conceptualized (Wutich et al., 2013). As Boelens, Dávila, and Menchú (1998) sum up, equity "deals with 'fairness in particular cases'" and is dynamically "formulated and functions in the communities themselves." As a result, equity is difficult to define universally and may be most fruitfully explored in a local context. To illustrate this further, Wutich et al. (2013) found that local conceptions of inequity are particularly salient in settings with water insecurity and inadequacy, suggesting dynamic interlinkages between conceptualization and operationalization of equity and the features of context—beyond cultural and institutional dimensions to also include biophysical and hydrological considerations.

A water-security capabilities approach to equity offers the opportunity to approach local conceptualizations of water (in)equities in ways that are both broadly understandable and locally specific. As mentioned previously, Goldin (2013, p. 315) argues that a capabilities approach to the water sector encompasses many dimensions, including human health and goods, education and literacy, significant relations with others, participation in social life, self-determination and autonomy, accomplishment, aspiration and self-respect, and basic mental and physical functionings. In this section, we address the ways that a water-security capabilities approach enables us to better understand different axes of inequity—socioeconomic, gender, and community—again emphasizing both the general and locally specific articulations of the concept. In relation to each of these dimensions, equity may take on different meaning and relevance—for instance, the ways that equity links with aspiration and self-respect or the ways that equity emerges as relevant to relations with others.

While it has been well-established that socioeconomic (in)equity is a core driver of water control, access, and use, there are a number of important (and as yet undertheorized) concerns related to equity that are potentially highlighted through a water-security capabilities framework. For instance, such an approach invites analysis of complex ways that socioeconomic inequities in hydro-social relations relate to broader capabilities and entitlements such as those associated with *education and literacy*. Educational exclusion, such as from information about water quality and water systems, is commonly a limited capability among politically marginalized communities. In South Africa, for example, racial and related economic inequities prevent residents of black and colored rural areas from obtaining the knowledge and power needed to confront historical water injustices. In the United States, residents of Flint, Michigan—who suffered lead poisoning due to municipal water mismanagement—were repeatedly prevented from obtaining timely and accurate information about lead risks (Katner et al., 2016). Although the historical, political, and economic dimensions of these cases are characterized by significant differences, conceptualizing education and literacy as a core capability enables us to better understand water security failures across both cases. Here, we see that there are key concerns not only with secure access to water but ways that inequities play into broader dynamics and capabilities (education, literacy), which impinge on water securities. As such, the reorientations bring into view other linked (structural, historical) inequities that are important for addressing and overcoming water insecurities and that remain obstacles to success in implementing the human right to water.

Adding to these understandings, a water-security capabilities approach enables us to theorize less-studied aspects of this phenomenon, including impacts on capabilities such as *significant relations with others* and *participation in social life*. For example, Sultana's (2011) work explores the complex social entanglements women in Bangladesh navigate as they attempt to obtain water that is uncontaminated by arsenic. Sultana demonstrates how women's relations with others are mobilized and compromised as they draw on attenuated family and neighborhood ties to access water. Wutich (2012) found that water crises in Cochabamba, Bolivia were moments in which both women and men were able to participate in social lives that transgressed normal gendered divisions of labor. In these cases, a water-security capabilities approach helps us think beyond received wisdoms about gender and water to look at a fuller range of capabilities and functionings. By attending to hydro-social relationships that are key to framing and supporting specific entitlements (such as entitlements to water), we can demonstrate a critical consideration that is often implicit, but somewhat underdeveloped, in water access, governance, and equity debates. To state this another way, we find a number of exciting connections are likely to be highlighted through a water security capabilities approach—key among them, we suggest, would be a fuller appreciation of social relationships, dynamics, and complexities that enable and condition uneven entitlements to water.

Across communities, water security also varies widely due to inequities in economic development, political power, and territorial sovereignty. Importantly, water (and water security) may be conceptualized in profoundly different ways across communities (Yates, Harris, & Wilson, 2017; Norman, 2017). Here, too, a water-security capabilities approach offers the possibilities for reconceptualizing, in culturally sensitive ways, capabilities such as *autonomy and self-determination* and *accomplishment*. The case of the San Francisco Peaks (*Nuvatukya'ovi*) conflict in the western United States illustrates this well. These mountains, located in Northern Arizona, have been the site of several legal battles over the use of land and recycled wastewater to support a ski resort (Glowacka, Washburn, & Richland, 2009; Schlosberg & Carruthers, 2010). The Hopi tribe recognizes the mountains as sacred home to ancestor deities; skiing, tourism development, and wastewater application all impede their religious practices and desecrate the site. As such, the recreational uses of the mountains impede the Hopi tribe's autonomy and self-determination. Local settlers value economic growth from the tourist economy and need the reclaimed wastewater to provide skiing opportunities to tourists in low-rain years. As such, the mountains offer these communities a sense of accomplishment, both in terms of income generated and skiing challenges. In this case, the water-security capabilities approach provides a framework for identifying, valuing, and evaluating different capabilities and functionings as they are conceptualized by local communities in conflict. Identifying these differentials as core to the conflict might serve as a key first step towards understanding the stakes and terms of the concerns at play.

Directions forward

Contemporary calls for a human right to water, regardless of how it is defined, pivots on a positive right: the ability of people to access a thing—in this case, water. What water is—its qualities, forms, conveyance system, or characteristics are debated—but this approach is moored to the water as a material object. The human right to water must be reframed on securing the relations that support people's and communities' relations with waterworlds as *inherent* aspects to what they choose to do and be. Access to water is clearly one element of this, but we argue, along with a growing number of scholars (Linton, 2012; Schmidt, 2012), that human dignity is coproduced through fluid relations. For these reasons, it is necessary to reconceptualize the human right to water in broad terms of the hydro-social relations (not only availability) that ensure human security, flourishing, and well-being—beyond merely water access or availability (Obeng-Odoom, 2012).

The water capabilities approach provides the opportunity for an important conceptual advancement on our understanding of the human right to water. The capability approach respects peoples' different ideas of the good life or in specific hydro-social relations in all its material and nonmaterial dimensions, and this is why *water-security capability* is the goal rather than any particular outcome. In doing so, the capability approach offers the possibility of focusing on hydro-social relations, such as water sharing rather than water as a material object. The HRW, as commonly conceptualized, focuses on water equality (ElDidi & Corbera, 2017; Stoler et al., 2018; Wutich et al., 2018; Brewis et al., 2019). However, the focus should be on water (in)equity. Equity is a much more complex concept than equality, difficult to define and implement outside of local contexts. Thus, a water capabilities approach provides a way to value and protect hydro-social terms of local needs, values, and ideas of a good life. For these reasons, adopting a CA approach to the human right to water could significantly advance its utility in accomplishing equity (not just equality). In this way, water security can operate within-against-and beyond the state in ways that promote the multiple hydro-social relations, and even differing water worlds, in support of human well-being.

Acknowledgments

We would like to thank Arizona State University's Center for Global Health and School of Human Evolution and Social Change for supporting the Household Water Insecurity Experiences (HWISE) workshop in which this chapter was developed. We would also like to thank Jessica Budds, Chad Staddon, Tennille Marley, Flavia Dantas, Roseanne Schuster, and Shalean Collins for insights on the chapter's direction during this meeting. Wendy Jepson would also like to thank Andreas Brannstrom for his editorial support in drafting the article.

References

Anand, P.B. (2007) *Scarcity, entitlements, and the economics of water in developing countries.* Cheltenham: Edward Elgar Publishing.

Angel, J. and Loftus, A. (2017) 'With-against-and-beyond the human right to water,' *Geoforum*, 98, pp. 206–213.

Bakker, K. (2007) 'The "commons" versus the "commodity": alter-globalization, anti-privatization, and the human right to water in the Global South,' *Antipode*, 39 (3), pp. 430–455.

Ballet, J., Marchand, L., Pelenc, J., and Vos, R. (2018) 'Capabilities, identity, aspirations and ecosystem services: An integrated framework,' *Ecological Economics*, 147, pp. 21–28.

Boelens, R., Dávila, G., and Menchú, R. (1998) *Searching for equity: conceptions of justice and equity in peasant irrigation.* Assen: Uitgeverij Van Gorcum.

Boelens, R., Perreault, T., and Vos, J. (2018) *Water justice.* Cambridge: Cambridge University Press.

Brewis, A., RosingerA., Wutich, A., Adams, E., Cronk, L., Pearson, A., Workman, C., Young, S., and the HWISE Consortium (2019) "Water sharing, reciprocity, and need: a comparative study of inter-household water transfers in Sub-Saharan Africa," *Economic Anthropology*, 6, pp. 208–221. Available at: https://doi.org/10.1002/sea2.12143

Bustamante, R., Crespo, C., and Walnycki, A. (2012) "Seeing through the concept of water as a human right in Bolivia," in Sultana, F. and Loftus, A. (eds) *The right to water: politics, governance, and social struggles*. New York: Routledge Publishing, pp. 223–240.

ElDidi, H. and Corbera, E. (2017). 'A moral economy of water: charity wells in Egypt's Nile Delta.' *Development and Change*, 48 (1), pp. 121–145.

Garrick, D. and Hall, J.W. (2014) "Water security and society: risks, metrics, and pathways," *Annual Review of Environment and Resources*, 39, pp. 611–639.

Gimelli, F.M., Bos, J.J., and Rogers, B.C. (2018) 'Fostering equity and wellbeing through water: A reinterpretation of the goal of securing access,' *World Development*, 104, pp. 1–9.

Glowacka, M., Washburn, D., and Richland, J. (2009) 'Nuvatukya'ovi, San Francisco Peaks: Balancing Western economies with Native American spiritualities,' *Current Anthropology*, 50 (4), pp. 547–561.

Goff, M. and Crow, B. (2014) 'What is water equity? The unfortunate consequences of a global focus on 'drinking water',' *Water International*, 39 (2), pp. 159–171.

Goldin, J. (2013) 'From vagueness to precision: raising the volume on social issues for the water sector,' *Water Policy*, 15 (2), pp. 309–324.

Grey, D. and Sadoff, C.W. (2007). 'Sink or swim? Water security for growth and development,' *Water Policy*, 9 (6), pp. 545–571.

Ingram, H., Whiteley, J.M., and Perry, R. (2008) "The importance of equity and the limits of efficiency in water resources," in Whiteley, J.A., Ingram, H., and Perry, R. (eds) *Water, place, and Equity*. Cambridge: MIT Press, pp. 1–31.

Jepson, W.E., Wutich, A., Colllins, S.M., Boateng, G.O., and Young, S.L. (2017) "Progress in household water insecurity metrics: a cross-disciplinary approach," *Wiley Interdisciplinary Reviews: Water*, 4 (3), pp. 1–21.

Jepson, W., Budds, J., Eichelberger, L., Harris, L., Norman, E., O'Reilly, K., Pearson, A., Shah, S., Shinn, J., and Staddon, C. (2017) 'Advancing human capabilities for water security: a relational approach,' *Water Security*, 1, pp. 46–52.

Jepson, W. and Vandewalle, E. (2016) 'Household water insecurity in the Global North: a study of rural and periurban settlements on the Texas–Mexico Border,' *The Professional Geographer*, 68 (1), pp. 66–81.

Katner, A., Pieper, K.J., Lambrinidou, Y., Brown, K., Hu, C.-Y., Mielke, H.W., and Edwards, M.A. (2016) 'Weaknesses in federal drinking water regulations and public health policies that impede lead poisoning prevention and environmental justice,' *Environmental Justice*, 9 (4), pp. 109–117.

Lankford, B., Bakker, K., Zeitoun, M., and Conway, D. (2013). *Water security: principles, perspectives, and practices*. New York: Routledge Publishing.

Lauderdale, P. (1998) 'Justice and equity: a critical perspective,' in Boelens, R. and Davila, G. (eds) *Searching for equity: conceptions of justice and equity in peasant irrigation*. Assen: Uitgeverij Van Gorcum, pp. 5–10.

Linton, J. (2010) *What is water? The history of a modern abstraction*. Vancouver, BC: UBC Press.

Linton, J. (2012) 'The human right to what? Water, rights, humans, and the relation of things,' in Sultana, F. and Loftus, A. (eds) *The right to water: politics, governance, and social struggles.* New York: Routledge, pp. 45–60.

Lu, F., Ocampo-Raeder, C., and Crow, B. (2014) "Equitable water governance: future directions in the understanding and analysis of water inequities in the Global South," *Water International*, 39 (2), pp. 129–142.

Meehan, K.M. (2014) 'Tool-power: water infrastructure as wellsprings of state power,' *Geoforum*, 57, pp. 215–224.

Mehta, L. (2014) 'Water and human development,' *World Development*, 59, pp. 59–69.

Mirosa, O. and Harris, L.M. (2012) 'Human right to water: contemporary challenges and contours of a global debate,' *Antipode*, 44 (3), pp. 932–949.

Morales, M. d. C., Harris, L., and Öberg, G. (2014) 'Citizenshit: the right to flush and the urban sanitation imaginary,' *Environment and Planning A*, 46 (12), pp. 2816–2833.

Morckel, V. (2017) 'Why the Flint, Michigan, USA Water Crisis is an urban planning failure,' *Cities*, 62, pp. 23–27.

Norman, E.S. (2017) 'Standing up for inherent rights: the role of indigenous-led activism in protecting sacred waters and ways of life,' *Society & Natural Resources*, 30 (4), pp. 537–553.

Nussbaum, M.C. (1997) 'Capabilities and Human Rights,' *Fordham L. Rev.*, 66, pp. 273.

Nussbaum, M.C. (2003) 'Capabilities as fundamental entitlements: Sen and social justice,' *Feminist Economics*, 9 (2–3), pp. 33–59.

Nussbaum, M.C. (2011a) "The capabilities approach and animal entitlements," in Beauchamp, T.L. and Frey, R.G. (eds) *The Oxford handbook of animal ethics*. Oxford: Oxford University Press, pp. 228–251.

Nussbaum, M.C. (2011b) *Creating capabilities.* Cambridge, MA: Harvard University Press.

Nussbaum, M. and Sen, A. (1993) *The quality of life.* Oxford: Oxford University Press.

Obeng-Odoom, F. (2012). 'Beyond access to water.' *Development in Practice*, 22 (8), pp. 1135–1146.

Pelenc, J. and Ballet, J. (2015) 'Strong sustainability, critical natural capital, and the capability approach,' *Ecological Economics*, 112, pp. 36–44.

Perera, V. (2015) 'Engaged universals and community economies: the (human) right to water in Colombia,' *Antipode*, 47 (1), pp. 197–215.

Perreault, T. (2014) 'What kind of governance for what kind of equity? Towards a theorization of justice in water governance,' *Water International*, 39 (2), pp. 233–245.

PopeFrancis (2017) "The human right to water." Dialogue on the Human Right to Water, Pontifical Academy of Sciences, The Vatican, February 24. Available at: https://press.vatican.va/content/salastampa/en/bollettino/pubblico/2017/02/24/170224a.html

Ranganathan, M. and Balazs, C. (2015) 'Water marginalization at the urban fringe: environmental justice and urban political ecology across the north–south divide,' *Urban Geography*, 36 (3), pp. 403–423.

Rauschmayer, F., Omann, I., and Frühmann, J. (2012) *Sustainable development: capabilities, needs, and well-being.* New York: Routledge.

Redfield, P. and Robins, S. (2016) 'An index of waste: humanitarian design, 'dignified living' and the politics of infrastructure in Cape Town,' *Anthropology Southern Africa*, 39 (2), pp. 145–162.

Robeyns, I. (2018) *Well-being, freedom, and social justice: the capability approach re-examined.* Cambridge, MA: Open Book.

Rodina, L. (2016) 'Human right to water in Khayelitsha, South Africa: lessons from a 'lived experiences' perspective,' *Geoforum*, 72, pp. 58–66.

Schlosberg, D. and Carruthers, D. (2010) 'Indigenous struggles, environmental justice, and community capabilities,' *Global Environmental Politics*, 10 (4), pp. 12–35.

Schmidt, J. (2012). 'Scarce or insecure? The right to water and the changing ethics of global water governance,' in Sultana, F. and Loftus, A. (eds) *The right to water: politics, governance, and social struggles*. New York: Routledge, pp. 94–109.

Scott, C.A., Meza, F.J., Varady, R.G., Tiessen, H., McEvoy, J., Garfin, G.M., Wilder, M., Farfán, L.M., Pablos, N.P., and Montaña, E. (2013). 'Water security and adaptive management in the arid Americas,' *Annals of the Association of American Geographers*, 103 (2), pp. 280–289.

Sen, A. (2000) 'Development as freedom,' *Development in Practice*, 10 (2), pp. 258–258.

Sen, A. (2005) 'Human rights and capabilities,' *Journal of Human Development*, 6 (2), pp. 151–166.

Shah, S.H., Angeles, L.C., and Harris, L.M. (2017) 'Worlding the intangibility of resilience: the case of rice farmers and water-related risk in the Philippines,' *World Development*, 98, pp. 400–412.

Stoler, J., Brewis, A., Harris, L.M., Wutich, A., Pearson, A.L., Rosinger, A., Schuster, R., and Young, S.L. (2018). "Household water sharing: a missing link in international health." *International Health*, 11 (3), pp. 163–165.

Sultana, F. (2011) 'Suffering for water, suffering from water: emotional geographies of resource access, control, and conflict,' *Geoforum*, 42 (2), pp. 163–172.

Sultana, F. (2018) 'Water justice: why it matters and how to achieve it,' *Water International*, 43 (4), pp. 483–493.

Sultana, F. and Loftus, A. (2012) *The right to water: politics, governance, and social struggles*. New York: Routledge.

Sultana, F. and Loftus, A. (2015) 'The human right to water: critiques and condition of possibility,' *Wiley Interdisciplinary Reviews: Water*, 2 (2), pp. 97–105.

United Nations High Commissioner for Refugees (2018) *Report of the Special Rapporteur on the issue of human rights obligations relating to the enjoyment of a safe, clean, healthy, and sustainable environment*. Available at: http://ap.ohchr.org/documents/dpage_e.aspx?si=A/HRC/34/49

Wutich, A. (2012) 'Gender, water scarcity, and the management of sustainability trade-offs in Cochabamba, Bolivia,' in McElwee, P. (ed) *Gender and sustainability: lessons from Asia and Latin America*. Tucson: University of Arizona Press, pp. 97.

Wutich, A., Brewis, A., Sigurdsson, S., Stotts, R., and York, A. (2013) 'Fairness and the human right to water,' in Wagner, J. (ed) *The social life of water*, New York: Berghahn Books, pp. 220–238.

Wutich, A., Budds, J., Jepson, W., Harris, L.M., Adams, E., Brewis, A., Cronk, L., DeMyers, C., Maes, K., Marley, T., and Miller, J. (2018) 'Household water sharing: a review of water gifts, exchanges, and transfers across cultures,' *Wiley Interdisciplinary Reviews-Water*, 5: e1309. Available at: https://doi.org/10.1002/wat2.1309

Yates, J.S., Harris, L.M., and Wilson, N.J. (2017) 'Multiple ontologies of water: politics, conflict, and implications for governance,' *Environment and Planning D: Society and Space*, 35 (5), pp. 797–815.

Zeitoun, M. (2011). 'The global web of national water security,' *Global Policy*, 2 (3), pp. 286–296.

Zeitoun, M., Lankford, B., Krueger, T., Forsyth, T., Carter, R., Hoekstra, A.Y., Taylor, R., Varis, O., Cleaver, F., Boelens, R., and Swatuk, L. (2016) 'Reductionist and integrative research approaches to complex water security policy challenges,' *Global Environmental Change*, 39, pp. 143–154.

Zwarteveen, M.Z. and Boelens, R. (2014) "Defining, researching, and struggling for water justice: some conceptual building blocks for research and action," *Water International*, 39 (2), pp. 143–158.

8 Rights on the edge of the city

Realizing of the right to water in informal settlements in Bolivia

Anna Walnycki

Reconciling state-sanctioned rights and local struggles for water

"There can be no human rights when the state is absent."[1]

"People say 'We have the right to water!' . . . but now what do we do, and what does it mean? There is no water here, the state has not brought it, so we have to do something ourselves."[2]

Special Rapporteurs for The Human Rights to Safe Drinking Water and Sanitation have documented the limitations of the right, specifically if national governments focus on achieving minimum obligations over more progressive goals (Albuquerque 2014). Meanwhile, the potential for informal service providers to contribute to progressive improvements to water services in cities that have urbanized without infrastructure has been documented (Bustamante, Crespo, and Walnycki 2012; Walnycki 2017; McGranahan et al. 2018), although grassroots organizations have voiced concerns about efforts to regulate informal water providers that might lead to political co-option (Olivera and Gomez 2006). Grassroots' concerns reflect the long history of exclusionary water governance practices in Latin America where certain groups exercise power over others (Castro 2008: 75), leading to conflict between the state and different groups, particularly when water services have been privatized or corporatized (Castro 2004). It is still unclear how a state-led, rights-based agenda can be used to improve water services in the poorest informal settlements, where state-citizen relationships are virtually nonexistent (see Mehta 2006).

When rights are associated with public provision, public-private binaries can become entrenched, obscuring the role and capacity of informal providers that play a crucial role in low-income urban settlements (Kooy 2014). Recognising the link and codependence between informal service providers to water utilities in many cities across the Global South presents new opportunities to improve access to water. In Latin America, the right to water has encouraged some governments to move beyond minimum core obligations and to pursue the progressive realization of the right. In cities that have urbanized without infrastructure, the unilateral utility model has been scrutinized and the potential of informal community providers to provide services is being recognized and supported (see Allen et al. 2017; Walnycki 2017).

Informal provision is sanctioned or supported through a range of ad hoc practices from tolerance to coproduction, which directly and indirectly sustain informal water providers, although this is rarely undertaken strategically and at scale. In Bolivia, reforms to develop policies and regulation around the right to water have sought to engage with the informal sector and pursue the progressive realization of the right, in principle at least. There are, however, enduring tensions and exertions of power between and within informal urban and rural stakeholders and the state, which continue to manifest as reforms to realise the right to water develop in Bolivia.

Bustamante, Crespo, and Walnycki (2012) set out the proposed realignment of state-citizen relationships through the regulation of all water services under the adoption of the human right to water. The struggle between state-sanctioned and community-won rights continues because the state presumed the authority to grant the right to water through new policies and laws. Although this is the norm in other countries, the state has never been able to achieve this in Bolivia, which reflects its long history of autonomous water resource and service management. Although this intervention is necessary for unserved urban settlements, where informal and community-based water services can be compromized by a host of physical, institutional, and economic challenges, many communities have resisted this exertion of state control, given the systematic exclusion of informal and low-income communities (Crespo 2010).

Rights are often incrementally recognized through local struggles, and despite efforts by the state to standardise the sector under banner of the right to water, informal water providers believe in a model that goes beyond minimum human rights premised on a fixed relationship between an individual and a quantity of water, where citizens play roles in the production of the hydro-social cycle and local water rights (Bustamante, Crespo, and Walnycki 2012), as expressed during the Cochabamba Water War.[3] Indeed, the political foundation of human rights-based approaches and their potential to achieve equality is premised on the active participation of citizens in processes of change as opposed to being inactive recipients of rights (May 2008: 143). Bolivian activists, social movement leaders, and community members who contest reforms around the right to water have done so because state-citizen relations have not served the poorest, who continue to rely on informal water services. Communities are concerned that reforms to realize the right to water and regulate informal collective water services might diminish their agency over service provision without guaranteeing improved services or fully recognising their rights as citizens (see Olivera and Gomez 2006).

Informality is the urban reality

In cities that have urbanized informally, water services are often delivered by a hybrid of interdependent informal and formal actors and institutions. Informal water providers are often embedded in urban service provision, making binaries between informal and formal basic service provision become increasingly misleading (Roy 2014). Informal services are operated by entrepreneurs for profit such as water trucks and groundwater supplies but can also be non-profit community responses to local needs, such as decentralized water systems or a community borehole that might receive practical and financial state support.

There have been numerous political efforts to incorporate indigenous and informal providers through policy and legal frameworks in Bolivia, reflecting the "institutional embeddedness" (Casson et al. 2010) of informal and indigenous institutions that have evolved over time. Laws and policies designed by the state, and informal norms to govern informal and indigenous services and markets, have come to reflect one another as an example of inter-legality (Regalsky 2009). In the absence of formal service provision, the state is often complicit in indirect practices that can maintain the informal sector. Tolerance, loose regulation, and fuzzy legislative frameworks are often present in settings where informal institutions have been allowed to flourish. This "calculated informality" (Roy 2009) fills service gaps, particularly for low-income groups that can't access formal services and markets but is rarely wholly supported by legislation and policies (see Kooy 2014). As a result, there are more risks and costs linked to services provided by the informal sector given that formal service providers are regulated, subsidized and/or overseen by the state.

There have been several specific attempts to reconfigure the relationship between formal and informal providers in Bolivia, most dramatically demonstrated during the Water War. Regulating informal water providers is another means by which the state has sought to broaden the base of institutions that can deliver basic services. However, if regulation doesn't meaningfully engage with informal providers, provide incentives and support, or is overly restrictive, it might discourage informal providers from adhering to reforms. This chapter now focuses on the efforts of collectively organized informal providers who have sought to reconfigure the distribution and governance of water services in Cochabamba, Bolivia, with emphasis on institutionalising equitable coproduced water services.

The empowering potential of coproduction partnerships has been of interest to urban social movements and researchers working in the Global South for some time. Allen et al. (2017) explore the different types of coproduction partnerships, or "platforms," that have emerged in marginalized urban communities around water and sanitation provision in Latin America. In instances where the state is committed to developing basic services with communities, coproduction initiatives have the potential to develop approaches that respond to the needs of communities but that draw on the technical expertise of municipal utilities (McGranahan 2015). These partnerships can enable communities to connect community infrastructure to formal urban infrastructure; for example, community-level sanitation can be linked up to the larger mainline pipes supplied by the state (ibid.).

Mitlin (2008) considers coproduction to be a political tool that is used by organized urban poor groups to engage with local governments, not only to improve access to basic services but also to develop a more strategic and advantageous relationship with the state that diffuses power to marginalized urban groups. While the state has often been framed as being the instigator of coproduction partnerships, Mitlin demonstrates how most community-based organizations exist despite the state. Coproduction is a method that is often adopted by community organizations to achieve scale, to shape development policies and politics around a pro-poor agenda, and even to challenge and transform the established structures of urban governance. Whether driven by the state or by grassroots processes, how informal organizations can influence depends on the quality of institutions (Casson et al. 2010). Allen et al. (2017) go a step further outlining the scope of coproduction partnerships to enhance political recognition and enhance the political capabilities of marginalized group in Latin America.

Water provision in Cochabamba: a patchwork of formal and informal providers

The following section considers how enduring models of informal community service provision have been shaped by political, social, and physical process in Cochabamba since the 1980s. It has been estimated that there are 500–600 small-scale formal and informal potable water providers across the municipality of Cochabamba (World Bank Water and Sanitation Programme 2007). Many are concentrated in the Zona Sur in the absence of any significant municipal provision.

Community water systems are particularly prevalent in older settlements and take diverse forms. Ledo (2013) geo-referenced around 200 independent systems in districts 7, 8, 9, and 14 of Cochabamba city. Most systems are managed by water committees (46 per cent), followed by water associations (20 per cent), formal neighborhood associations sanctioned by decentralization and known as territorial base organizations (OTBs, 15 per cent) and cooperatives (11 per cent). Meanwhile, the Metropolitan Master Plan (MMAYA 2013) identified 189 small local systems managed by OTBs (23 cases), self-management (122), small cooperatives (11), private urbanizations (26), and agrarian *sindicatos*.[4] These community water providers provide according to communitarian principles and have been shaped to varying degrees by the various national policy, community, and physical process over time.

Community water services in Cochabamba are built using community resources and labor and deliver the service via household connections. The systems rely on groundwater or water deliveries, from the utility of private vendor to a community storage tank. Households pay for water used and for maintenance and are expected to contribute to the upkeep of the system. The cost of water varies from community to community and can be more or less expensive than SEMAPA water, depending on investment in infrastructure and availability of a groundwater source. Principle challenges, beyond securing access to safe affordable water sources when groundwater sources have become increasingly saline and depleted, include building a sustainable management model and maintaining community participation (see Walnycki 2013).

Fuzzy laws, emerging rights, and tenacious community organizations

Bolivia's first (and only) water law was passed in 1906. The law recognized water as a public good, but in practice, access to water was based on private rights to water that were generally tied to property and land rights. Landlords with water on their land had usufruct rights if it had no direct negative impact on third parties. The MNR (Revolutionary Nationalist Movement) government instigated reforms that would lead to the 1967 constitution that recognized all Bolivians over the age of 18 as citizens, regardless of their status, property ownership, or occupation. The constitution was the culmination of a series of political reforms that led to the dissolution of the *haciendas*, the abolition of landlords and the redistribution of land to peasants. The reforms during this period sought to replace indigenous identity with ideas of class, embodying a shift towards the assimilation of indigenous communities (Albro 2010: 74) and state-citizen rights and obligations. The reforms failed to homogenise the population; indeed, it had quite the opposite effect, entrenching indigenous and communal forms of organization (Dunkerley 2007: 74).

Strong state institutions did not replace the landowner, meaningful state-citizen relationships were not forged. Communal organizations such as agrarian unions and indigenous councils persisted and grew. These events influenced the formation of the organizational structures and institutions that shape water provision (irrigation and potable) in Bolivia to this day. The 1953 Ley de Reforma Agraria (Law of Agrarian Reform), which dissolved the haciendas also brought de facto water rights to previously marginalized peasant communities. This affected how water was managed and distributed in the north and northwest of Cochabamba. Communities of peasant farmers developed communally managed irrigation systems, which were often formed around preexisting irrigation systems of old *haciendas*. These irrigation systems are based on dynamic ideas of uses and customs that were shaped by the physical constraints of the river basin, historic and cultural process, mutually agreed water rights, and the long-term management of the river (Perredo et al. 2003: 11). This law would influence the development of unions of collective water providers, and this model came to influence the development of other associations of communal water providers in peri-urban areas.

Poor city planning thwarted by informal urbanization

Around the same time as the political reforms of 1953, the architect Urqidi took on the sociopolitical project of modernising Cochabamba. Cochabamba would be transformed from a rural market town to a contemporary city, through a planning process that drew heavily on the ideas of the European "garden city." Urquidi attempted to zonify modern Cochabamba and delineate its development through greenbelts that would be preserved for agriculture and leisure. The orderly development of this garden city failed to respond to three distinctive waves of migration that followed.

The first wave followed the end of bonded labor in the late 1940s (Kohl et al. 2011). The city was unable to provide enough affordable housing for the growing population, and so small, sporadic informal settlements began to emerge in the arid, mountainous land to the south of the city, which had previously been only sparsely populated by some farmers. Urban drinking water providers emerged in the informal communities that were starting to be consolidated around Cochabamba in the late 1950s. So-called "land invasions" (Goldstein 2004: 72) continued throughout the 1960s and 1970s, then during the 1980s, after the global crash in the cost of metals and the closure and privatization of the mines in accordance with a structural readjustment programme, a third wave of mining migrants led to the development of new informal communities but, this time, further south of the municipality.

Housing and basic services in the Zona Sur developed largely because of the impetus of "savvy" *loteadoras* [5] trying to make a quick buck, sustained by the absence of any real regulation of the expansion. Remittances and profits from the coca-growing funded the development of landlord-led housing in many communities. The state overlooked the invaders and *loteadoras* until the early 1990s, creating a peri-urban population that was socially, economically, and politically marginalized from the outset. During the 1970s, the municipality initiated a policy of exclusion, illegalization, and criminalization of the informal *barrios* that were emerging on the peri-urban fringes of the city. By 1993, 80 per cent of all the peripheral *barrios* were categorized as clandestine or illegal (ibid.: 72), thus, entrenching the role of informal water service providers.

There has been a citywide water provider in Cochabamba since 1948, but the public water utility SEMAPA was established in its first incarnation in 1967 as a decentralized public body responsible for the technical, administrative provision of water and sanitation across the municipality of Cochabamba. The municipalization of water provision during this period did not extend to the whole municipality. Mirroring the trend observed in many developing countries during the 20th century, Cochabamba experienced an infrastructural crisis reflecting broader economic and political processes (Laurie and Marvin 1999). The municipality became more densely populated and simultaneously continued to expand, urbanizing the surrounding regions, without developing sufficient infrastructure to meet the water and sanitation needs of the city, particularly for the poorer migrant communities in the peri-urban Zona Sur. The public utility, like the sector at large, was underfunded until the 1990s, with less than 1 per cent of public investment in the sector prior to privatization (Oporto and Salinas 2007) and with infrastructure being prioritized for the urban centre, meaning that the Zona Sur relied on informal communal provision and water vendors.

Decentralization legitimises informal community providers

Bolivia followed its Latin American neighbors by adopting decentralization and participation practices during the 1990s. The 1994 Law of Popular Participation (LPP) sought to decentralize the influence of the state by taking advantage of preexisting, informal, communitarian, or social structures, including indigenous councils (*ayllus*), in rural areas and neighborhood associations in urban areas. Communities could legalize their neighborhood to form an OTB (Organisación Territorial del Base or grassroots community organization). These institutions could access per capita funds for local development initiatives and participate in decentralized decision-making processes at the local council, district, and municipality. However, the process was not accompanied with the same resources, autonomy, or innovations such as participatory budgeting that were seen in Brazilian, Uruguayan, and Colombian cities. Instead of creating space for meaningful participation, decentralizing reforms were creating spaces for nonstate actors and the private sector to deliver services. Furthermore, reforms were interpreted as an approach intended to build local groups, in order to reduce the influence of unions and social organizations (Arbona 2007: 28) by appropriating ideas of multiculturalism that were linked to broader homogenising neoliberal policies adopted across Latin America during this period (McNeish and Lazar 2006).

Although popular participation had some success adapting to rural indigenous and mining community institutions, the same effective participatory processes were not mirrored in urban areas. OTBs operating in this region have been tainted by corruption and elite capture, meaning that development processes have been ineffective and piecemeal (see Torrico and Walnycki 2015). Furthermore, many communities continue to be "informal" and unrecognized by the state. In early 2015, 39 of the 73 communities in district 8 in the Zona Sur of Cochabamba were informal and unrecognized by the municipality and could not access decentralized resources (ibid.).

Some OTBs used some LPP funding to develop community water systems, demonstrating that there has been indirect financial support for community water services since the 1990s, further entrenching the role of communities in basic service delivery in the region. Parallel "informal" community water committees or associations persisted and were tolerated by the state with the support of the church, pooled community funds, mutual aid, and community labor. They endured given the shortcomings of the urban utility, further diminishing lines of responsibility and accountability between citizens and the state around service provision.

The ripple effects of the Water War

There have been several grassroots processes and social movements that have shaped relations between informal water providers and the Bolivian state and that have had varying degrees of success, the most well documented of which was the 2000 Water War. These events led to the recognition of urban community water providers by the state, as part of wider efforts to standardise provision in a manner that recognises and respects traditional forms, rights, uses, and customs under the banner of the human right to water (MMAYA 2008). The narrative that developed around the Water War stated that community water providers across the Zona Sur of Cochabamba were part of the uprising. Many of the water committees from the southern zone were not part of the uprising.[6]

Community water providers benefitted from the influx of resources and support from international NGOs immediately after the Water War, and it became politically unviable for the state to undermine the autonomy of community water providers in rural and urban areas across Bolivia. These processes underpinned the policy shift that followed the election of Evo Morales as president in 2006, which in principle suggested that the state would pursue a more inclusive model of water governance that would incorporate and develop informal providers. Closer analysis shows that many of these reforms do not mark a radical departure from past policies, and there are significant questions around the extent to which they can improve water services for low-income urban neighborhoods.

Reforming the sector around the right to water

The human right to water and sanitation was enshrined in the Bolivian constitution that was approved in 2009 as part of a spectrum of human and indigenous rights. The right to water enabled the Bolivian state to assume responsibility for guaranteeing the right. Water and sanitation provision can be provided through public utilities at the municipal level, cooperatives and community providers such as drinking water committees or mixed entities, collectively known as water and sanitation service providers (EPSAS). In 2008, the Bolivian state estimated that there were over 28,000 largely informal, small-scale providers nationally (SENASBA 2012). Institutional reforms to the sector are based on the constitutional commitment to the reform and regulation of formal and informal water providers. The Ministry for Water and the Environment (MMAYA) was established in 2008 to coordinate the sector; the Authority for the Fiscalisation of and Social Control over Drinking Water and Sanitation (AAPS) oversees regulation; the National Service for the Sustainability of Basic Sanitation Services (SENASBA) is responsible for ensuring the sustainability of all water providers through community development strategies and technical assistance; and the Environment and Water Executing Agency (EMAGUA) and the National Productive and Social Investment Fund (FPS) oversee the implementation of programmes and projects formulated by the MMAYA. The MMAYA along with institutions such as SENASBA and AAPS are in place to support small-scale water providers to be able to deliver adequate water services. To date, the institutional groundwork for these reforms has been undertaken and the MMAYA received a US$ 78 million-dollar loan from the Inter-American Development Bank to develop and consolidate an inclusive model that supports community water providers as an interim solution to provision (MMAYA 2012).

The proposal to recognise and include informal water providers as service providers builds on a sector-approach that pre-dates the election of Evo Morales. Perreault (2008) and Marston (2014) allude to the significance of Law 2066, which was introduced prior to the Water War and granted community and indigenous organizations the right to apply for the exclusive right to grant water services in specific regions. This was concretized by the establishment of the Technical Committee of Licenses and Registrations (CTRL) in 2009, which was put in place to register and grant licences, although few of the estimated 28,000 informal community providers have pursued this approach. Even with a licence for provision, the state has been slow to develop an appropriate regulatory framework to engage with small-scale water providers.

The shortcomings in the formal Bolivian water sector can all be broadly attributed to underinvestment, poor regulatory frameworks, or inefficient implementation of regulation. AAPS has, thus, been tasked with establishing a decentralized structure of regulation that incorporates small EPSAS, consolidating their role and their relationship to the state.

This horizontal approach to regulation has the scope to support and build the capacity of informal providers that are so central to water service provision, but it has some way to go. Thousands of community providers continue to operate informally, and the poorest households tend to rely on unregulated mobile vendors. Furthermore, the state has limited knowledge and access to the informal sector, and this is further complicated by limited resources and capacity to engage with the providers that make up this fragmented water sector. This reinforces the unequal forms of formal and informal provision that characterise the sector and further embed the pluri-legal nature of water governance arrangements in Bolivia.

Grassroots struggles shape rights

Some community water providers have federated around grassroots processes designed to secure funding and practical support, to address some of the technical and institutional challenges facing small-scale water providers. They have also gained the political recognition necessary to influence city-level water governance arrangements and the development of national water policies and institutions.

A noteworthy example of networked grassroots organizing among urban community water providers is that of the Association of Communitarian Water Systems and EPSAS of the Zona Sur and the department of Cochabamba (ASCIASUDD–EPSAS). The post-Water War political climate provided community water providers with new sources of finance and support from NGOs that moved into the region. They were also protected by the state, following the introduction of Law 2066 and the continued political lobbying around the rights of community providers by the Coordinadora del Agua that led the Water War. Until this point, the CWPs had always operated independently and often in competition for access to water sources, but following the Water War, SEMAPA and civil society organizations began to consider the practical scope of CWPs as alternative water providers. The Social Committee for Life was established to promote coordination between committees, to strengthen DWCs as institutions, and to create an umbrella organization so that the DWCs of the Zona Sur could interact with SEMAPA. This process was developed by NGOs, activists, the church, and SEMAPA, and it reflected some of the discourses and demands that had arisen out of the Water War. These included social control, protecting and galvanising communal water providers, and developing new forms of public water management that incorporated community and citizen participation (Grandiddyer 2006).

The platform provided an impetus for communities to form their own network of urban CWPs that sought to reform water governance arrangements around a model of *co-gestion* or co-management (Grandiddyer 2006: 246). ASCIASUDD–EPSAS sought to engage with the reforms that were ushered in following the election of Evo Morales. At a national level, they participated in discussions around the right to water in the constitution and in consultations around a proposed new water law. The efforts developed from a position around the failure of decentralization and popular participation in the low-income *barrios* of the Zona Sur (Grandiddyer 2006: 350).

This approach yielded some fruit during the early years of the Morales presidency, as ASICASUDD–EPSAS received some decentralized finance to distribute to water committees to upgrade community systems. The EU-funded PASAAS project provided a major source of finance for DWCs belonging to ASICASUDD–EPSAS to develop or upgrade their infrastructure for provision. In practice, funds were distributed between 41 DWCs for the construction of tanks, for upgrading and extending pipelines, and for the installation of sewage systems. Most projects were completed by the end of 2011, resulting in 8,280 connections and benefiting 41,400 people in the Zona Sur (ASICASUDD–EPSAS 2012). The organization also successfully promoted the development of two decentralized pipelines that could deliver water to member water committees, upon completion of the Misicuni dam project.[7] These proposals continue to be central to plans under development by the municipal authority as a means of improving access to water for unserved low-income settlements.

Despite the practical and strategic gains made by ASICASUDD–EPSAS, by 2012, the federation had disbanded due to internal institutional disputes that arose as political and financial state support declined. In subsequent years, CWPs have received some technical support from NGOs working with municipal authorities to improve water service provision on a committee-by-committee basis, but there is less space for low-income groups to shape discussions around water management and governance.

How to progressively reorder state-citizen relationships around rights

Reforms to realize the right to water in Bolivia are the latest attempt to reorder relationships between the state and informal providers, building on decentralization during the 1990s and ongoing efforts to consolidate national water laws and policies. This approach to water governance has often been described by urban and rural social movements, who are engaged in the collective production of water services in low-income urban settlements, as an exclusionary exercise of state power. Early reforms around the right to water had the potential to support a more progressive approach that recognized and supported informal providers to play a role in incrementally improving service provision, given the ongoing physical and political challenges to improving water services in the region.

The human right to water and global goals such as SDG6 define an increasingly depoliticized arena occupied by international agencies and institutions that are focused on finance, technologies, water quantities, quality, and issues of access at scale. Global rights and goals do not easily support the informal efforts and struggles of low-income urban communities that work to identify and respond to the local practical and political water needs of men, women and children, through vending, water committees, decentralized water systems, advocacy and lobbying, and efforts to co-produce basic services with utilities.

The potential and progress made by informal providers are rarely documented or recognized, let alone supported by national policies, or integrated into plans to improve water services in urban settlements. Although provision continues to be delivered by co-dependent formal and informal providers in cities like Cochabamba, sector finance and governance arrangements rarely recognise or adequately support informal providers. National policies that enable utilities to provide strategic support and engage with organized communities such as the water committees in Cochabamba provided a glimpse of how marginalized communities could be engaged in the production and governance of water services. Further technical support, funding, and strategic decision making should be devolved to local groups and processes to improve provision and enhance the state-citizen relationships required for the right to water to be progressively recognized.

Notes

1 Participant in a workshop on the right to water, Cumbre por el Agua y el Saneamiento Basico, Cochabamba, 26 June 2010.
2 President of the Water Committee, Cochabamba, 30 May 2012.
3 During the mid-1990s, Bolivia came under pressure from the World Bank to privatize water service provision in Cochabamba, as a precondition of debt relief packages from the World Bank and the International Monetary Fund. In 1999, the Cochabamba contract was awarded to the sole bidder, Bechtel, which was granted exclusive rights both to provide water services and to access all the water sources in Cochabamba, including the aquifer in the region. The concession included plans to develop the Misicuni dam for hydropower, irrigation, and potable water on the outskirts of

the city. The cost of the dam, combined with a contractual ruling against public subsidies to protect customers against price hikes, meant that bills increased by 200 per cent in some communities within two months of the concession being granted. Numerous urban and rural communities that relied on local water sources rose in opposition to the concession and the appropriation of resources, and these protests were to develop into the Water War. This urban–rural alliance became known as the Coordinadora del Agua and eventually forced the annulment of the concession in 2001. Cochabambinos earned themselves an international reputation as the world's "water warriors" and the Water War would come to play a central role in anti-privatization and right to water narratives in subsequent years.

4 Agrarian *sindicatos* (peasant unions) are organizational structures created after the Agrarian Reform of 1953 and constitute the maximum authority at rural community level. They are formally recognized as a legal organization and represent the inhabitants in each community. Everybody who owns land in a community must affiliate to the *sindicato*. Many local issues, including local water management for irrigation and domestic purposes, are governed through the *sindicato* (Cossío et al. 2010: 6), and it is through this organization that each community establishes contact with state and nonstate organizations. Cochabamba has some agricultural land in districts 8 and 9 where *sindicatos* persist.

5 *Loteadoras* would illegally appropriate unused land on the edge of the city, carve it into blocks and sell it off to migrants who wanted to build cheap housing in the region.

6 Water secretary, Cochabamba, 10 September 2011

7 The Misicuni multipurpose project was first proposed in 1952 and has been through several incarnations since. The dam will channel water from the Misicuni river to the Cochabamba valley and will supposedly provide sufficient potable water for all seven metropolitan municipalities in the valley, irrigation water for local farmers, and has an electricity component. Work on the project was halted in 2013 because of contractual disputes, and the project has since been marred by controversy.

References

Albro, R. (2010) Confounding cultural citizenship and constitutional reform in Bolivia. *Latin American Perspectives*, 37 (3), pp. 71–90.

Albuquerque, C. (2014) *Realising the human rights to water and sanitation: A handbook*. Geneva: OHCHR.

Allen, A., Walnycki, A., and von Bertrab, É. (2017) "The co-production of water justice in Latin American cities," in Allen, A., Griffin, L., and Johnson, C. (eds) *Environmental justice and resilience in the urban Global South*. New York: Palgrave McMillan, pp. 175–193.

Arbona, J. (2007) Histories and memories in the organization and struggles of the Santiago II neighbourhood of El Alto, Bolivia, *Bulletin of Latin American Research*, 27 (1), pp. 24–42.

ASICASUDD–EPSAS (2012) "Proyecto PASAAS." Available at: www.asicasuddepsas. org/contenidos/leer/25-05-2011-PROYECTO-PASAAS

Bustamante, R., Crespo, C., and Walnycki, A. (2012) "Seeing through the concept of water as a human right in Bolivia," in Sultana, F. and Loftus, A. (eds) *The right to water: Politics, governance, and social struggles*. London and New York: Routledge, pp. 223–240.

Casson, M.C., Della Giusta, M., and Kambhampati, U.S. (2010) "Formal and informal institutions and development," *World Development*, 38 (2), pp. 137–141.

Castro, J.E. (2004) Urban water and the politics of citizenship: The case of the Mexico City Metropolitan Area during the 1980s and 1990s, *Environment and Planning A*, 36 (2), pp. 327–346.

Castro, J.E. (2008) "Water struggles, citizenship, and governance in Latin America," *Development*, 51 (1), pp. 72–76.

Cossio, V.et al . 2010*Cooperation and conflict in local water management: Conflict and cooperation in local water governance – inventory of local water-related events in Tiraque district, Bolivia(DIIS Working Paper 11)*. Copenhagen: Danish Institute for International Studies

Crespo, C. (2010) *El derecho humano en la practica: La politica del agua y los RRNN de gobierno Evo Morales*. CESU–UMSS Ponencia al Congreso de Sociologia. Bolivia: Centro de Estudios Superiores Universitarios de la Universidad Mayor de San Simón.

Dunkerley, J. (2007) Evo Morales, the 'two Bolivias' and the third Bolivian revolution, *Journal of Latin American Studies*, 39 (1), pp. 133–166.

Goldstein, D.M. (2004) *The spectacular city: Violence and performance in urban Bolivia*. Durham: Duke University Press Books.

Grandiddyer, A. (2006) La lucha por el agua en Cochabamba: Manifiesto de los hombres y mujeres de la Zona Sur, Cochabamba, Marzo de 2006, in Gutiérrez, R. and Escárzaga, R. *Movimiento Indígena en América Latina: Resistencia y Proyecto Alternativo*. Volumen II. pp. 68–74.

Kohl, B., Farthing, L.C., and Muruchi, F. (2011) *From the mines to the streets: A Bolivian activist's Life*. University of Texas Press.

Kooy, M. (2014). Developing informality: The production of Jakarta's urban waterscape, *Water Alternatives*, 7 (1), pp. 35–53.

Laurie, N. and Marvin, S. (1999) "Globalisation, neoliberalism, and negotiated development in the Andes: Bolivian water and the Misicuni dream," *Environment and Planning A*, 31 (8), pp. 1401–1415.

Ledo, C. (2013) *El agua nuestra de cada día. Retos e iniciativas de una Cochabamba incluyente y solidaria*. Cochabamba: CEPLAG–UMSS.

Marston, A.J. (2014) The scale of informality: Community-run water systems in peri-urban Cochabamba, Bolivia, *Water Alternatives*, 7 (1), pp. 72–88.

May, T. (2008) *The political thought of Jacques Ranciere: Creating equality*. Edinburg, PA: Edinburgh University Press.

McGranahan, G. (2015). Realizing the right to sanitation in deprived urban communities: Meeting the challenges of collective action, coproduction, affordability, and housing tenure, *World Development*, 68, pp. 242–253.

McGranahan, G., Walnycki, A., and Dominick, F., Kombe, W., Kyessi, A., Limbumba, T.M., Magambo, H., Mkanga, M., and Ndezi, T. (2018) How international water and sanitation monitoring fails deprived urban dwellers, in Cumming, O. and Slaymaker, T. (eds). *Equality in Water and Sanitation Services*. Abingdon: Routledge.

McNeish, J. and Lazar, S. (2006) The millions return: Democracy in Bolivia at the start of the 21st Century, *Bulletin of Latin American Research*, 25 (2), pp. 157–162.

Mehta, L. (2006) *Water and human development: Capabilities, entitlements and power* (Human Development Report Office Occasional Paper 8). Available at: http://hdr.undp.org/sites/default/files/hdro_1303_stewart.pdf

Ministerio de Medio Ambiente y Agua (2008) *Plan Nacional de Saneamiento Básico 2008-2015 (Ministerio de Medio Ambiente y Agua)*. La Paz, Bolivia: Ministerio de Medio Ambiente y Agua.

Ministerio de Medio Ambiente y Agua (2012) *Loan proposal for Inter-American Development Bank: Reform programme for the water and sanitation and water resources sectors in Bolivia*. Bolivia: Ministerio de Medio Ambiente y Agua.

Ministerio de Medio Ambiente y Agua (2013) *Plan maestro metropolitano de agua potable y saneamiento del área metropolitano de Cochabamba*. Bolivia: Ministerio de Medio.

Mitlin, D. (2008) With and beyond the state: Coproduction as a route to political influence, power, and transformation for grassroots organizations, *Environment and Urbanization*, 20 (2), pp. 339–360.

National Service for the Sustainability of Basic Sanitation Services (SENASBA) (2012) Available at: http://senasba.gob.bo

Olivera, O. and Gomez, L. (2006) *The rising waters*. Ottawa, Canada: The Blue Planet Project.

Oporto, H. and Salinas, L. (2007) *Agua y Poder. Fundación Milenio and Centre for International Private Enterprise*. La Paz: Imprenta Creativa.

Peredo, C. et al. (2003) *Los regantes de Cochabamba en la guerra del agua: presión social y negociación*. Bolivia: Centro de Estudios Superiores Universitarios de la Universidad Mayor de San Simón.

Perreault, T. (2008) Custom and contradiction: Rural water governance and the politics of usos y costumbres in Bolivia's irrigators' movement, *Annals of the Association of American Geographers*, 98 (4), pp. 834–854.

Regalsky, P. (2009) *Indigenous territoriality and decentralisation in Bolivia (1994–2003): Autonomies, municipalities, social differentiation and access to resources*. Newcastle: University of Newcastle Upon Tyne.

Roy, A. (2009) Why India cannot plan its cities: Informality, insurgence, and the idiom of urbanization, *Planning Theory*, 8 (1), pp. 76–87.

Roy, A. (2014) *Capitalism: A ghost story*. London: Verso.

Torrico, M. and Walnycki, A. (2015) Las Chompas en el poder? El mito de la participación en los barrios pobres de Cochabamba, Bolivia, *Medio Ambiente y Urbanización*, 82 (1), pp. 81–116.

Walnycki, A.M. (2013) *Rights on the edge: The right to water and the peri-urban drinking water committees of Cochabamba*. Doctoral dissertation, University of Sussex.

Walnycki, A.M. (2017) "Contesting and coproducing the right to water in peri-urban Bolivia," in Bell, S., Allen, A., Hofmann, P., and Teh, T.-H. (eds). *Urban water Trajectories* (Vol. 6). London: Springer, pp. 133–147.

World Bank Water and Sanitation Programme (2007). *Estudio sobre Operadores Locales de Pequeña Escala en Áreas Periurbanas de Bolivia*. La Paz, Bolivia: World Bank.

9 Human right to water and bottled water consumption

Governing at the intersection of water justice, rights and ethics

Raúl Pacheco-Vega

Introduction

Bottled water is one of the world's top businesses even in countries where water stress is acute. The global market is valued at about 250 billion USD, and it's only expected to grow. Despite the popularity of sugary, fizzy drinks, bottled water is now the top choice for individual hydration. This fact is inherently paradoxical. It's hard to justify the extraction of thousands of liters of the vital liquid often from depleted aquifers, while populations in water-stressed regions suffer from lack of access and inequity in distribution. Just as an example, while Mexico is considered a highly water-stressed country, it is the biggest global consumer per capita of water in containers with approximately 154 to 254 liters per person per year (Pacheco-Vega, 2015a; 2015b). Even though Mexican consumers usually purchase from 20-liter containers for their households on a twice-a-week basis, single-serving usage has also grown substantially (Pacheco-Vega, 2017).

Individuals decide to consume bottled water and a majority of studies point out to a fear of the tap and lack of trust in local infrastructure drives this, although in affluent societies it may also be the result of developing a taste for it (Saylor, Prokopy & Amberg, 2011; McLennan, 2015; Viscusi, Huber & Bell, 2015). Some people consume bottled water because they believe that it's a healthier option for their hydration purposes. Strong marketing and powerful branding strategies (Brei, 2018) promote bottled water as the drink of choice for healthy hydration (Race, 2012; Hawkins, Potter & Race, 2015), all the while promoting a notion of purity that is associated with consuming the packaged liquid (Opel, 1999; Geissler & Gamble, 2002; Wilk, 2006; Sharma & Bhaduri, 2014). Though clearly there's a rising taste for consuming packaged water in the form of one liter or 500 ml portable plastic containers (Biro, 2017), individuals across many countries have indicated their consumption of bottled water is driven primarily by distrust in government, and particularly, in the robustness and appropriate maintenance of water-delivery infrastructure (Johnstone and Serret, 2012). Consumption of the individually-packaged liquid is, thus, used as a reverse-quarantine mechanism (Szasz, 2007) whereby each person takes care of his or her own safety through the application of physical barriers that protect their consumptive practices. For many people, it's a consumer lifestyle choice (York et al., 2011), but for many others, it's the only way to protect their health. For some African countries, consuming the packaged liquid in sachets is a unique mechanism to enact the human right to water (Stoler, 2012; Morinville, 2017).

Water is an important public policy problem partly because of its uneven and often inequitable global availability and wide-ranging variability in distribution and consumption. While ownership of the resource (and, therefore, its governance) is allocated differently across countries, regulatory bodies frequently task adequate water supply to subnational, primarily local governments. Nevertheless, given the nature of water as a scarce, common pool resource, how can governments across countries and governance levels ensure that the human right to water as enacted by the United Nations' 2010 Resolution can really be implemented? This is a key policy challenge that we can't overlook and necessitates discussion, particularly in contexts where water scarcity is high, and the operation of bottled water companies is, thus, perceived as unjust. It's this paradox that I discuss in this paper: how can bottled water exist in a context of water insecurity, and what are the moral, ethical and policy implications of its continued existence? How can we govern bottled water at the intersection of water insecurity, ethics and justice?

In this chapter, I explore how bottled water production and consumption complicates the challenge that implementing the human right to water posits. I examine the literature on the human right to water, justice, ethics and the security/insecurity continuum to discuss whether it is ethical to operate a packaging-water business in regions facing high degrees of water stress and whether individuals perceive this operation as just or ethically correct. I argue that implementing the human right to water in a context of extensive bottled water consumption is fraught with challenges. The discussion here centers on the role of governments in guaranteeing that their citizens can make use of their human right to water in a context where bottling companies have taken a central role in delivering water within urban and rural contexts. How can we ensure that a governmental function is not supplanted by a private actor? How do we guarantee that citizens can access water within their means when this vital liquid is widely commodified? And to what extent and under what circumstances can or should we allow bottled water companies to take up the role that ought to be the state's? I argue that allowing bottling companies to temporarily help with water provision can be ethically and morally advisable but that this intervention should be temporary. Otherwise, we face a real risk that communities that could be using tap water or seeking ways to reduce their bottled water consumption may go back and further commodification of the vital liquid will continue, to the detriment of potentially robust implementation of the human right to water.

This paper adds to the small-but-growing literature on the politics of bottled water and to the well-established field of study of the human right to water by discussing rights to water in the context of packaged liquid consumption. I do not examine the politics of soft drinks and eschew a discussion of the politics of beer. I do focus on exploring the morality of packaged water production and consumption in collapsed contexts where disaster strikes, and water is scarce. My objective is to exemplify the complex relationship that packaged water has to fulfill the human right to water. My critique centers on the ways in which multinational and domestic bottling companies use public relations campaigns that portray them as "good Samaritans" when their business is to market packaged water. I also discuss how citizens who supply disaster-stricken communities with donations of bottled water could inadvertently end up promoting it as a permanent solution to a public policy problem instead of a temporary relief measure, thus, leading to a dereliction of duty on the part of governments, who will not see it as their duty to ensure that all citizens are able to have universal water access.

Throughout the chapter, I use vignettes from two case studies to discuss instances where bottled water production and consumption could be considered "ethical" or at least, valid: the usage of bottled water during the recent Mexico City earthquake and a short-lived campaign by Anheuser-Busch and MillerCoors to stop producing beer and start manufacturing canned water for use during Hurricane Harvey. These vignettes serve to highlight the inherent paradox in the existence of packaged water within a context of extreme water scarcity and challenging engineering, infrastructure and governance issues that hinder the enactment of the human right to water. I argue that these socially responsible actions obscure the reality that these companies effectively commodify water, making it challenging for communities within to enjoy access to water.

The chapter is structured as follows: in the second section, after this introduction, I offer a political economy view of bottled water. I discuss how bottled water can be viewed as an example of accumulation by dispossession. In the third section, I discuss the ethics of producing and consuming bottled water through a water injustice framework. In the fourth section, I briefly describe the vignettes. Using these, I analyze issues of temporality, fetishization, dramatization and spectacle and water injustice, to conclude in the fifth section with a summary of the core argument.

Accumulation by dispossession and water as a political and economic resource: a political-economy examination of bottled water

Given water's characteristics as a common pool resource (or commons), transforming it into a commodity that is extracted, traded and distributed across the globe posits a puzzle: how can governments ensure that every individual at the subnational level can have access to enough of the vital liquid to survive and thrive, without blocking industrial production of packaged water, as some activists appear to be asking for?[1] While the human right to water has been increasingly predicated as a global norm that every country ought to achieve, there is still more than a billion people who lack access to safe and sufficient sources of water. Governments do struggle to create and maintain infrastructure to provide water across their governed territory and communities have variegated geographies and contexts. Implementing the right to water in one city may not be as easy as doing it in another. Therefore, we ought to take a cautious approach to impose the right to water as a framework (Sultana and Loftus, 2015) for urban, rural and peri-urban water delivery and instead have a more nuanced discussion regarding how states can go about it.

Using water as a political resource has been common practice for centuries. While there are many ways in which water is weaponized (Del Giacco et al., 2017), one of the most important is controlling access and availability of the vital liquid. Denying access to water to vulnerable communities has been done to force residents to ensure their cooperation in upholding and enforcing specific policies or simply to shift their behavioral patterns towards those that are more amenable for governments in power. Just as an example, Cleaver documents how water access was denied by the state in the Nkayi District, in Zimbabwe, in order to bring residents to support unpalatable fiscal and economic and settlement policies posited by the government (Cleaver, 1995).

Manipulation and control of communities using water resources can be done through various methods, including restricting access, controlling flow and quality and negating consumption. Restricting access can be done on a temporary or permanent basis. Flow control is usually done to force individual households to pay their water utility bills. Quality control is often dependent on having the right treatment process and maintaining operating installed capacity. Restricting access (i.e. negating consumption) can be achieved through infrastructure modifications (e.g. shutting down entire pipelines) whereas denying consumption can be achieved through the refusal of infrastructure provision (e.g. de-prioritizing urgent connection to water and sewerage networks in budget allocation exercises or preventing access to water pipes).

Water marketization is one of the strategies through which rent is extracted from this vital natural resource. I follow a similar typology to that presented in Pacheco-Vega (2015a), based on works by Leila Harris and Karen Bakker. I see privatization, marketization and commodification as three examples of ways in which markets interact with water. Privatization, the partial or total transfer of operations from a public water utility to a private entity, is considered a mechanism to encourage utilities to become more efficient and reduce expenditures, to provide better service for communities and offer continuous service at adequate pressure and flux. Marketization is the creation of markets associated with the value of water. One of the ways in which water is marketized is through the emergence of formal and informal water markets, where access to the vital liquid is associated with a buyers' and sellers' market. While not packaging the resource, marketizing does assign a monetary value and, therefore, creates a transactional process and, thus, a market, be it formal or informal. Finally, the commodification of water involves its productification: that is, making of water a product with characteristics that allow it to be sold within a market. When water is commodified, it is usually packaged and sold. Water being treated as a merchandise is commodified water.

By either constitutional mandate or bylaws, many subnational governments are in charge of providing safe drinking water (Pacheco-Vega, 2018). Regardless of whether this responsibility is provincial (state- or regional-level) or municipal (city-metropolitan area), it is a public duty. Publicness theory asserts that the duty of public servants and public entities is to provide public services (Osborne, Radnor &Nasi, 2013; Talmage, Anderson & Searle, 2018). While it is perfectly normal, and often frequently done, that public service delivery can be undertaken by private entities (Bakker, 2010; Furlong & Bakker, 2010; Furlong, 2012; 2016), the continued provision of a public service by a private entity, in particular one where the service is an actual commodity, a packaged product that is sold at an exorbitant price, becomes both an ethical issue and a public policy one. On the public policy side, governments that delegate the responsibility for accessing good quality drinking water are also abdicating their responsibility to serve the public. And on the ethical side, having governments push communities for the purchase of a commodity at a high mark-up to satisfy communities' needs for drinking water becomes, therefore, rather troubling if not directly unethical. I expand on this discussion in the following sections.

The ethics of bottled water consumption and production: a water injustice framework

Bottled water has been criticized for marketing using dubious safety and health-related claims (Dupont, (Vic) Adamowicz & Krupnick, 2010; McLennan, 2015; Ragusa & Crampton, 2016). For example, Nestlé, through their Gerber brand, sells "water made for your baby". Danone, through its Mexican brand Bonafont, promotes "light water", which allegedly will make whoever drinks it feel lighter. Another valid critique has to do with excessive production of polyethylene (PET) plastic bottles, which will afterward end up in landfills (Royte, 2008; Gleick, 2010; Hawkins, 2011). Other criticisms include the fact that often times, huge amounts of water are being extracted unregulated at a ridiculously low cost for the company, leaving communities negatively affected (Tisdale, 2004; Jaffee & Newman, 2013a; Jaffee & Case, 2018). Whether it is rational for individuals to consume bottled water because of a distrust of their tap water (Parag & Roberts, 2009; McSpirit & Reid, 2011; Viscusi, Huber & Bell, 2015) is also debated because it depends on consumers' experience with bottled water and tap water. This distrust may be warranted in locations where infrastructure for urban water delivery through municipally-owned utilities is poor (Anand, 2012; Ranganathan, 2016; Prasetiawan, Nastiti & Muntalif, 2017). In these cases, consuming bottled water does become a rational choice. Yet other critiques involve a more philosophical discussion of how bottling companies are commodifying the vital liquid (Snitow, Kaufman & Fox, 2007; Melosi, 2011; Jones, Murray & Overton, 2017) and concerns for the potential loss of income of local water utilities. Spending money on a commodity could weaken a municipal water utility as consumers choose to purchase bottled water, which could be equated to potential revenue loss (Pacheco-Vega, 2017).

Branding packaged water as "ethical" implicitly associates a cultural meaning of propriety and appropriateness, of implicit justice and morality. But there is nothing implicitly just about extracting water and appropriating rent through the marketization of the vital liquid. There are at least three ways in which bottled water can be framed as "ethical" and branded as such (Potter, 2011; Brei & Böhm, 2014; Hawkins & Emel, 2014). First, bottled water production is framed as ethical when its manufacturing process reduces its impact on the environment. One example is Boxed Water, the beverage that is produced by packaging water using cardboard and other "natural" materials. Second, packaged water consumption is framed as ethical when it enables vulnerable or marginalized populations to access the resource, and where there would be no other alternatives (Vedachalam et al., 2017). The third one would be through programs that promote the consumption of ethically-branded bottled water to support charitable purposes. Starbucks' Ethos brand is allegedly "ethical bottled water" because when a customer purchases a bottle, Starbucks donates 5 cents to a program that supports water and sanitation projects in developing countries.

Brei and Böhm (2011) criticize marketing of "ethical" bottled water as an example of a branding strategy where corporate social responsibility is wielded through the signification of symbolic messages about the inherent ethics of consuming bottled water that is promoted as "just" or "responsible". A similar critique could be wielded against allegedly socially responsible brands providing free packaged water to citizens affected by disasters. By framing canned water and donated water as conduits for the human right to water, as mechanisms to ensure that afflicted populations have access to the vital liquid, we are also implicitly framing them as ethically-packaged water. Producing and consuming packaged water has potentially negative implications for regions with high water stress or depleted aquifers. Therefore, it's important to discuss the ethical dilemmas associated with the global bottled water industry and its impact on local scales.

While there are different versions of the concept of "water justice", particularly that espoused by Zwarteveen and Boelens (2014), less has been written about the notion of "water injustice". Perhaps one of the latest incarnations of the discussion has emerged when describing the emergence of water utility marketization forces across many countries. Perceived and procedural justice are equally important, as indicated by these authors,

> "[u]nderstanding (in)justice, then, encompasses the examination of both formally accredited justice (formal schemes of interpretation and legitimization, and legal-positivist constructs of 'rightness') and socially perceived justice or equity (location-, time-, and group-specific constructs of 'fairness') that are used by different societal groups."
>
> (Zwarteveen and Boelens, 2014, p. 147)

Water injustice is, thus, perceived as the creation of inequities through privatization, marketization and commodification processes (Pacheco-Vega, 2017).

Extracting water from regions where water is already scarce, marketing it as a healthy alternative to municipally-provided water and profiting from what should be a commons, thereby making it a commodity, all complicate the relationship between the existence of bottled water and fulfilling the human right to water. Continued profiteering from water extraction and packaging by multinational corporations can potentially lead to water scarcity and poverty, an unjust activity if we view it through political ecology perspective (Loftus, 2009).

One of the most important critiques of water injustice comes from the human right to water movement (Sultana and Loftus, 2015). While Sultana and Loftus do an excellent job of clarifying why the human right to water can't be a direct policy tool, thus requiring that we engage with the possibility that the concept itself has only been wielded by politicians and has become not only politicized but also potentially an empty signifier, in Sultana and Loftus' exact words (p. 98), it is still used as a policy directive. But the question of whether local governments are obligated to enact the human right to water has remained in the global discourse, particularly because issues of who will pay for access when specific, vulnerable populations are unable to access the vital liquid will always arise in a market economy. This discussion remains unresolved but serves to highlight the importance of considering the complex relationship between the existence of bottled water and fulfilling the human right to water goal.

Discussions around water injustice provoked by bottled water are most frequently associated with negative perceptions towards bottling companies (Pacheco-Vega, 2016; Jaffee & Case, 2018) which extract water and profit from a resource that in theory should be available to all humans. This is a valid critique at a time when water is becoming ever scarcer. These critiques are equally valid as well in the case of "ethically-branded bottled water" because the notion of ethical bottled water is a strange paradox: how can something that represents the epitome of David Harvey's "accumulation by dispossession" have any ethical component (DeChaine, 2012)? Bottling water and producing beer are industrial activities that can be perceived as unjust because their mere existence is based on a process of "accumulation by dispossession" (Jaffee & Newman, 2013b; 2013a; Walsh, 2015). While companies add value through extracting and packaging these liquids, they, at the same time, extract rent from communities and reduce access to their own resources.

To summarize, my argument is as follows: while bottled water exists as a global industry, consuming the vital liquid in packaged form also creates challenges for the attainment of the human right to water. It would follow that it is unethical to consume bottled water (a commodity) in contexts where depleted aquifers and surface water bodies like lakes and rivers are the norm. Nevertheless, there are instances where consuming packaged water is necessary, for example when the quality of networked, piped water supply is not high, as is the case in many urban and rural regions in Mexico, Indonesia and other countries. So, in those instances, consuming bottled water would be considered ethical and for governments, providing it would be the responsible thing to do. Not every city is equipped to provide safe drinking water, and in some cases, the delivery of packaged water is what enables and facilitates the enjoyment of the human right to water. But this ought to be a temporary process, rather than a permanent one. When governments abdicate their responsibility to provide safe drinking water to all citizens and instead allow private entities such as multinational corporations to profit from communities' inability to access drinking water at the household level, we witness a commodification of the human right to water. Furthermore, even when providing bottled water to disaster-stricken areas is ethically responsible, these companies should still remain under scrutiny, so as to ensure that their water extraction is sustainable and not negatively affecting communities. I expand this argument further in the next few pages.

Packaged water production and consumption in disaster vs. everyday contexts

Mexicans top the per-capita ranks of bottled water consumption globally with 194 liters per day, increasingly growing in the past few years (Pacheco-Vega, 2015a). Though scattered and not very well organized, there have been some discussions on ways to reduce wastefulness and generation of plastic residues in landfills within the activist community and governmental actors. The government of Mexico City enacted a law that required restaurants to offer filtered tap water instead of forcing customers to buy bottles of water. Even though these efforts seem to have had some effect on the consumption of bottled water, other pyrrhic victories have contributed to an increase in its production, as packaged water acts as a substitute for soda pop in anti-obesity campaigns. Nevertheless, Mexico City's anti-bottled water law appeared to be having a positive effect in reducing consumption of the packaged liquid. That is until the 19S earthquake hit. On September 19, 2017, a 7.3 intensity earthquake shook Mexico City, leaving thousands of people homeless and (by government counts) about 150 people deceased. This earthquake, eerily replicating the one from 1985, had a much less negative impact than its predecessor, but nevertheless, left much of the city's vulnerable water network infrastructure damaged, thus leaving thousands of people unable to access water not only for drinking purposes but also for general survival necessities. Xochimilco, the small city-within-a-lake, with its chinampas and touristic centers, took a very heavy toll. During reconstruction efforts, and over the next few days after the earthquake, a lot of people started donating entire cases of bottled water. Huge mountains of bottles of the vital liquid were showcased across social media platforms as humanitarian gestures and demonstrations of "just how much people in Mexico cared for their own people".

Hurricane Harvey hit Houston from August 25 through 29, 2017, leaving devastation across the middle and upper Texas coast.[2] Though Harvey hit many vulnerable communities, its impact on the city of Houston made quite an impression worldwide because the city was perceived to be almost invulnerable to disasters, and the disaster showcased how its urban development was so poorly planned. During the weeks following Harvey's massive destruction, several major beer companies have begun producing canned water through a change in their production systems to help affected communities access the vital liquid.[3] Two of the most visible and popular efforts were Anheuser-Busch Brewery in the state of Georgia, and MillerCoors in Chicago, Illinois. According to reports, Anheuser-Busch had more than 50,000 cans delivered to Louisiana and sent 100,000 to Texas, whereas MillerCoors sent 50,000 to the Red Cross in Texas, along with a monetary donation of $25,000.[4] The major reason underlying the usage of canned water was the complete devastation of water utility infrastructure.

Table 9.1 summarizes the examples discussed here through a lens of dramatization and fetishization. As I argue here, there's a "superhero" fetishization and dramatization of packaged water as THE response to a natural disaster. In both cases, individual actors are praised: society at large in the case of providing free bottled water to communities afflicted by the September 19, 2017, Mexico City earthquake and two beer production companies for temporarily producing canned water to give away to those affected by Hurricane Harvey in Houston, in the United States. This praising of individual actors for facilitating access to packaged water has an important implication for a discussion on the human right to water. Framing bottled water production as an ethical and just activity is problematic, in the same way as the consumption of free, donated packaged water was predicated as the only way in which these populations would be able to consume the vital liquid. The spectacle that this created in the Mexico City case had the unintended effect of promoting bottled water as a mechanism to ensure that citizens were able to enjoy their right to water. This promotional effect built goodwill for bottled water companies that would otherwise be heavily scrutinized.

Situations of crisis create spaces for manipulation and control of resources and communities. Because access to drinking water is vital for populations afflicted by disasters, there is little, if any, local opposition to acquisition and distribution of bottled or canned water. Ensuring that ravaged areas can have enough water for their survival needs then becomes a priority, and if using packaged water is the mechanism to enact the human right to water, producing and distributing it is also then considered justifiable. Nevertheless, it's important to ensure that these corporate social responsibility practices don't end up

Table 9.1 Two vignettes that exemplify private actors substituting public action towards enabling the human right to water

	USA	Mexico
Natural disaster	Hurricane Harvey	19S earthquake
Packaged water strategy	Halting beer production in Colorado	Donating thousands of bottles of water
"Superhero"	Anheuser-Busch and MillerCoors	Society at large

becoming promotional and legitimizing tools for the continued exploitation and commodification of water resources.

Who chooses the communities and the temporal horizon of where and when can canned or bottled water be freely distributed? How can we ensure that these allocation and selection processes are fair? These are all valid questions that necessitate further probing. While the corporate citizenship initiative that these companies are engaging in can be truly and honestly the result of genuine interest in communities' well-being, it's important that we carefully consider the implications of these temporal and spatial constraints on enacting the human right to water using packaged liquids. How much longer, from the date in which disaster struck, will a beer company continue to provide free canned water? What kind of decision processes will its stakeholders engage? Will discontinuation of water provision leave afflicted communities in worse condition than they were before? This is an important consideration to make because households can become complacent/confident that they will continue to have free water provision when this is only a temporary measure.

Ensuring access to water is a governmental function, not a corporate one. Therefore, water utilities should not rely on companies' good corporate citizenship to be able to provide a service that they are supposed to offer. Offering free bottled water to disaster-stricken populations wrongly implies that delivering the human right to water through providing a commodity for free on a temporary basis can legitimately become a policy tool on a long-term basis.

A Foucauldian argument would posit that enabling the human right to water using packaged water can easily become a mechanism for biocontrol (Collier, 2009). Who gets to decide who accesses emergency packaged water, and can they control decision-making processes at the individual level through holding people hostage to the consumption (or lack there-of) of bottled or canned water? This question should concern water scholars, as water can also be commodified and simultaneously weaponized.[5] Sultana is very clear and correct in her assessment: water is about power, and about how this power is harnessed and wielded.

> Water is essentially about power – the power to decide, control, allocate, manage – thereby affecting people's lives. This is intersectionally experienced by gender, class, race, and other axes of social difference, therefore affecting different groups of people in varied ways. Isolating a specific water issue thus often misses out on broader connections that tie peoples, places, policies and ecologies in far-flung places.
>
> (Sultana, 2018, p. 485)

There is a clear justification by citizens in Mexico City and Houston and their neighboring areas that purchasing large quantities of bottled water and donating them to resource centers for redistribution across the city and other affected areas is not only reasonable but even more so, necessary. The temporary commodification of water as a resource becomes not only justifiable but an imperative. The temporality of production and consumption of packaged water in a context of disaster relief is the key factor to consider here. While it is true that these corporate social responsibility efforts are laudable, and one should be happy that beer companies donate packaged water when disaster strikes, it is also necessary to evaluate whether we would agree with this approach if the circumstances were different, or whether it should continue for an extended period beyond reconstruction efforts. This discussion necessitates an assessment of how long the process of reconstruction would take and how much packaged water would be necessary to continue relief efforts.

Conclusions: what are the implications of the human right to water in the case of bottled water consumption?

Bottled water exists as a commodity because it responds to the specific needs of communities and individuals who, for one reason or another, are either unable or unwilling to access high-quality potable drinking water through either piped networks or large containers for redistribution/refilling (20 liters, in most cases). As I've argued throughout the chapter, a discussion on the ethics and morality of bottled water as a vehicle to enact the human right to water is complex, because it reaches beyond the simple discussion of whether one should be packaging the vital liquid and selling it. Extracting water from one region to ensure access in another facing stress brings up issues of legitimacy, water ownership, property and usage rights. Who should be able to access water within a certain geographical region and what are the limits and scale at which a certain population should be denied access to a water body?

There are several challenges facing governments who are committed to ensuring that all households can make their right to water valid in a context of increasing commercialization of the vital liquid. In this paper, I have outlined the challenges that negating bottled water consumption can bring when a certain population has endured disasters such as hurricanes, earthquakes or other types of abrupt climatic events. While rejecting the commodification of water, we are also confronted with the necessity of considering temporary relief measures that justify bottled water consumption. However, I argue that we ought to ensure that these efforts are temporary, and we should also be able to set a time horizon within which we resume conversations and considerations about the risks to ecosystem health and human security that marketization of the vital liquid presents.

One of the biggest problems of engaging in this discussion while overlooking the realities of needing to supply packaged water to communities afflicted by disasters is that the conversation becomes one-sided: "one should not be extracting the vital liquid from depleted aquifers for profit". This is where temporality, as I briefly discuss throughout the text, is important. The "superhero" solution is necessarily temporary. Governments ought to do their job, which is to ensure that individuals are able to enjoy their human right to water. Yes, it would be ethical to temporarily provide packaged water to communities and individuals under duress, but not at the expense of assigning responsibility to local water utilities. At the same time, governments ought to strengthen human capital and infrastructure if they are committed to providing drinking water as a public service (Pigeon, 2012).

At the same time, temporary solutions are only that: supposed to exist only for a defined and definite period. But lack of access to water and more generally, the inability to engage in full enjoyment of the human right to water can lead individuals and communities to seek other solutions, particularly accessing informal water markets (Venkatachalam, 2015; Wutich, Beresford & Carvajal, 2016). Wutich and collaborators, with their analysis of Bolivian informal water vending and Venkatachalam with a study in India, showed how these informal solutions often become more formalized as time goes by. The temporality issue arises again: the human right to water is not supposed to be a fleeting, volatile human right. One is supposed to be able to enjoy it regularly, even if its contours and conditions of possibility are complex (Schmidt, 2012; Sultana & Loftus, 2012; 2015). Adapting to water insecurity conditions through creating or accessing informal water markets distorts the very purpose of governments' responsibilities to provide safe drinking water. In a neoliberal context that has been pushing for slimmer, shallower governments, it is imperative that we ensure that the human right to water is not superseded by private, informal and individual-based solutions, but collective ones. The latter are both the purpose and the responsibility of governments. We should not let them off the hook.[6]

Notes

1 Several anti-bottled water activists have clearly stated the importance of preventing bottling companies from either obtaining new licenses or increasing water extraction with relative success. For example, note how activists in the city of Guelph, in Ontario, Canada, successfully lobbied the provincial government to call for a moratorium on new and expanded bottling operations. See: https://news.ontario.ca/ene/en/2016/10/ontario-taking-action-to-protect-clean-water.html
2 See: www.weather.gov/crp/hurricane_harvey
3 See: www.nydailynews.com/news/national/beer-companies-switch-canned-water-harvey-victims-article-1.3452701
4 See: https://theknow.denverpost.com/2017/08/29/oskar-blues-miller-anheuser-water-canning-2017/156917

   I am grateful to Dr. Annie Sugar for suggesting ways in which I could use the morality of beer production and consumption as one of my case studies.
5 While in this chapter I make biocontrol arguments, I do not seek to use a Foucauldian framework throughout. I am more interested in showcasing the importance of considering water as a means of control. As Sultana (2018) reminds us very clearly, water is about power.
6 I am grateful to the editors of this volume for extremely insightful comments that helped me improve the chapter. I am also thankful to Luis Alberto Hernandez Alba for superb research assistantship, participants in the 2018 American Association of Geographers panel, Christiana Zenner, Andrew Biro, Katie Meehan, Alida Cantor, Wendy Jepson, Amber Wutich, Kate Parizeau, Nina Gallagher, Kate O'Neill also provided excellent comments that also improved the argument I present here.

References

Anand, N. (2012) 'Municipal disconnect: On abject water and its urban infrastructures', *Ethnography*, 13 (4), pp. 487–509.
Bakker, K. (2010) *Privatizing water: Governance failure and the world's urban water crisis*. Ithaca, NY: Cornell University Press.
Biro, A. (2017) "Reading a water menu: Bottled water and the cultivation of taste", *Journal of Consumer Culture*, pp. 1–21. Available at: https://journals.sagepub.com/doi/10.1177/1469540517717779
Brei, V.A. (2018) 'How is a bottled water market created?', *Wiley Interdisciplinary Reviews: Water*, 5 (1), p. e1220.
Brei, V. and Böhm, S. (2011) 'Corporate social responsibility as cultural meaning management: A critique of the marketing of 'ethical' bottled water', *Business Ethics: A European Review*, 20 (3), pp. 233–252.
Brei, V. and Böhm, S. (2014) '"1L=10L for Africa": Corporate social responsibility and the transformation of bottled water into a 'consumer activist' commodity', *Discourse & Society*, 25 (1), pp. 3–31.
Cleaver, F. (1995) 'Water as a weapon: The history of water supply development in Nkayi District, Zimbabwe', *Environment and History*, 1 (3), pp. 313–333.
Collier, S.J. (2009) 'Topologies of power', *Theory, Culture & Society*, 26 (6), pp. 78–108.
DeChaine, D.R. (2012) 'Ethos in a bottle: Corporate social responsibility and humanitarian Doxa', in Dingo, R.A. and Scott, J.B. (eds) *The megarhetorics of global development*. Pittsburgh, PA: University of Pittsburgh Press, pp. 75–100.
Del Giacco, L.J., Lucentini, L. and Murtas, S. (2017) "Water as a weapon in ancient times: Considerations of technical and ethical aspects", *Water Science and Technology: Water Supply*, 17 (5), pp. 1490–1498.
Dupont, D., (Vic) Adamowicz, W.L. and Krupnick, A. (2010) 'Differences in water consumption choices in Canada: The role of socio-demographics, experiences and perceptions of health risks', *Journal of Water and Health*, 8 (4), pp. 671–686.
Furlong, K. (2012) 'Good water governance without good urban governance? Regulation, service delivery models and local government', *Environment and Planning A: Economy and Space*, 44 (11), pp. 2721–2741.
Furlong, K. (2016) *Leaky governance: Alternative service delivery and the myth of water utility independence*. Vancouver, BC: UBC Press.
Furlong, K. and Bakker, K. (2010) 'The contradictions in 'alternative' service delivery: Governance, business models and sustainability in municipal water supply', *Environment and Planning C: Government and Policy*, 28 (2), pp. 349–368.
Geissler, G.L. and Gamble, J.E. (2002) 'Straight from the tap?', *Journal of Food Products Marketing*, 8 (2), pp. 19–32.
Gleick, P.H. (2010) *Bottled and sold: The story behind our obsession with bottled water*. Washington, DC: Island Press.
Hawkins, G. (2011) 'Packaging water: Plastic bottles as market and public devices', *Economy and Society*, 40 (4), pp. 534–552.
Hawkins, G., Potter, E. and Race, K. (2015) *Plastic water: The social and material life of bottled water*. Boston, MA: The MIT Press.
Hawkins, R. and Emel, J. (2014) 'Paradoxes of ethically branded bottled water: Constituting the solution to the world water crisis', *Cultural Geographies*, 21 (4), pp. 727–743.
Jaffee, D. and Case, R.A. (2018) 'Draining us dry: Scarcity discourses in contention over bottled water extraction', *Local Environment*, 23 (4), pp. 485–501.

Jaffee, D. and Newman, S. (2013a) 'A bottle half empty: Bottled water, commodification and contestation', *Organization & Environment*, 26 (3), pp. 318–335.

Jaffee, D. and Newman, S. (2013b) 'A more perfect commodity: Bottled water, global accumulation and local contestation', *Rural Sociology*, 78 (1), pp. 1–28.

Johnstone, N. and Serret, Y. (2012) 'Determinants of bottled and purified water consumption: Results based on an OECD survey', *Water Policy*, 14 (4), pp. 668–679.

Jones, C., Murray, W.E. and Overton, J. (2017) 'FIJI Water, water everywhere: Global brands and democratic and social injustice', *Asia Pacific Viewpoint*, 58 (1), pp. 112–123.

Loftus, A. (2009) 'Rethinking political ecologies of water', *Third World Quarterly*, 30 (5), pp. 953–968.

McLennan, J.D. (2015) 'Choosing bottled over tapped: Drinking water in the Dominican Republic', *Journal of Water, Sanitation and Hygiene for Development*, 5 (1), pp. 9–16.

McSpirit, S. and Reid, C. (2011) "Residents' perceptions of tap water and decisions to purchase bottled water: A survey analysis from the Appalachian, Big Sandy coal mining region of West Virginia", *Society & Natural Resources*, 24 (5), pp. 511–520.

Melosi, M.V. (2011) *Precious commodity: Providing water for America's cities*. Pittsburgh, PA: University of Pittsburgh Press.

Morinville, C. (2017) 'Sachet water: Regulation and implications for access and equity in Accra, Ghana', *Wiley Interdisciplinary Reviews: Water*, 4 (6), p. e1244.

Opel, A. (1999) 'Constructing purity: Bottled water and the commodification of nature', *The Journal of American Culture*, 22 (4), pp. 67–76.

Osborne, S.P., Radnor, Z. and Nasi, G. (2013) 'A new theory for public service management? Toward a (public) service-dominant approach', *The American Review of Public Administration*, 43 (2), pp. 135–158.

Pacheco-Vega, R. (2015a) 'Agua embotellada en México: De la privatización del suministro a la mercantilización de los recursos hídricos', *Espiral: Estudios sobre Estado y Sociedad*, 22 (63), pp. 221–263.

Pacheco-Vega, R. (2015b) "Urban wastewater governance in Latin America: Panorama and reflections for a research agenda", in Aguilar-Barajas, I., Mahlknecht, J., Kaledin, J., Kjellen, M. and Mejia-Betancourt, A. (eds) *Water and cities in Latin America: Challenges for Latin America*. London: Earthscan/Taylor and Francis, pp. 102–108.

Pacheco-Vega, R. (2016) 'The global politics of bottled water: Towards a research agenda', in *2016 Meeting of the International Studies Association (ISA)*. Atlanta, GA: International Studies Association, pp. 1–16.

Pacheco-Vega, R. (2017) "Agua embotellada en Mexico: Realidades, retos y perspectivas", in Pacheco-Vega, R., Denzin, C. and Taboada, F. (eds) *El agua en México: Actores, sectores y paradigmas para una transformación social-ecológica*. Ciudad de México, México: Friedrich Ebert Stiftung, pp. 195–214.

Pacheco-Vega, R. (2018) 'Policy styles in Mexico: Still muddling through centralized bureaucracy, not yet through the democratic transition', in Howlett, M. and Tosun, J. (eds) *Policy styles and policy-making: Exploring the linkages*. Abingdon and New York: Routledge, pp. 89–112.

Parag, Y. and Roberts, J.T. (2009) 'A battle against the bottles: Building, claiming and regaining tap-water trustworthiness', *Society & Natural Resources*, 22 (7), pp. 625–636.

Pigeon, M. (2012) "Who takes the risks?", in Pigeon, M., McDonald, D.A., Hoedeman, O. and Kishimoto, S. (eds) *Remunicipalisation: Putting water back into public hands*. Amsterdam, Netherlands: The Transnational Institute, pp. 74–89.

Potter, E. (2011) 'Drinking to live: The work of ethically branded bottled water', in Lewis, T. and Potter, E. (eds) *Ethical consumption: A critical introduction*. Oxon: Routledge, pp. 116–130.

Prasetiawan, T., Nastiti, A. and Muntalif, B.S. (2017) '"Bad" piped water and other perceptual drivers of bottled water consumption in Indonesia', *Wiley Interdisciplinary Reviews: Water*, 4 (4), p. e1219.

Race, K. (2012) '"Frequent sipping": Bottled water, the will to health and the subject of hydration', *Body & Society*, 18 (3–4), pp. 72–98.

Ragusa, A.T. and Crampton, A. (2016) 'To buy or not to buy? Perceptions of bottled drinking water in Australia and New Zealand', *Human Ecology*, 44 (5), pp. 565–576.

Ranganathan, M. (2016) 'Thinking with Flint: Racial liberalism and the roots of an American water tragedy', *Capitalism Nature Socialism*, 27 (3), pp. 17–33.

Royte, E. (2008) *Bottlemania: Big business, local springs and the battle over America's drinking water*. New York: Bloomsbury USA.

Saylor, A., Prokopy, L.S. and Amberg, S. (2011) 'What's wrong with the tap? Examining perceptions of tap water and bottled water at Purdue University', *Environmental Management*, 48 (3), pp. 588–601.

Schmidt, J.J. (2012) 'Scarce or insecure? The right to water and the ethics of global water governance', in Sultana, F. and Loftus, A. (eds) *The right to water: Politics, governance and social struggles*. London and New York: Earthscan, pp. 94–109.

Sharma, A. and Bhaduri, S. (2014) 'Consumption conundrum of bottled water in India: An STS perspective', *Bulletin of Science, Technology & Society*, 33 (5–6), pp. 172–181.

Snitow, A., Kaufman, D. and Fox, M. (2007) *Thirst: Fighting the corporate theft of our water*. New Jersey, NJ: Josey-Bass.

Stoler, J. (2012) 'Improved but unsustainable: Accounting for sachet water in post-2015 goals for global safe water', *Tropical Medicine & International Health*, 17 (12), pp. 1506–1508.

Sultana, F. (2018) 'Water justice: Why it matters and how to achieve it', *Water International*, 43 (4), pp. 483–493.

Sultana, F. and Loftus, A.J. (eds) (2012) *The right to water: Politics, governance and social struggles*. London and New York: Earthscan.

Sultana, F. and Loftus, A. (2015) 'The human right to water: Critiques and condition of possibility', *Wiley Interdisciplinary Reviews: Water*, 2 (2), pp. 97–105.

Szasz, A. (2007) *Shopping our way to safety: How we changed from protecting the environment to protecting ourselves*. Minneapolis: University of Minnesota Press.

Talmage, C.A., Anderson, D.M. and Searle, M.S. (2018) 'Whither recreation and parks? Understanding change in public institutions through a theory of adaptive publicness', *Perspectives on Public Management and Governance*, 1 (2), pp. 143–158.

Tisdale, M. (2004) 'The price of thirst: The trend towards the privatization of water and its effect on private water rights', *Suffolk University Law Review*, 37 (1), pp. 535–553.

Vedachalam, S., MacDonald, L.H., Omoluabi, E., OlaOlorun, F., Otupiri, E. and Schwab, K.J. (2017) "The role of packaged water in meeting global targets on improved water access", *Journal of Water Sanitation and Hygiene for Development*, 7 (3), pp. 369–377.

Venkatachalam, L. (2015) 'Informal water markets and willingness to pay for water: A case study of the urban poor in Chennai City, India', *International Journal of Water Resources Development*, 31 (1), pp. 134–145.

Viscusi, W.K., Huber, J. and Bell, J. (2015) 'The private rationality of bottled water drinking', *Contemporary Economic Policy*, 33 (3), pp. 450–467.

Walsh, C. (2015) 'Mineral springs, primitive accumulation and the "new water" in Mexico', *Regions and Cohesions*, 5 (1), pp. 1–25.

Wilk, R. (2006) 'Bottled water: The pure commodity in the age of branding', *Journal of Consumer Culture*, 6 (3), pp. 303–325.

Wutich, A., Beresford, M. and Carvajal, C. (2016) "Can informal water vendors deliver on the promise of a human right to water? Results from Cochabamba, Bolivia", *World Development*, 79, pp. 14–24.

York, A.M., Barnett, A., Wutich, A. and Beatrice, C. (2011) "Household bottled water consumption in Phoenix: A lifestyle choice", *Water International*, 36 (6), pp. 708–718.

Zwarteveen, M.Z. and Boelens, R. (2014) 'Defining, researching and struggling for water justice: Some conceptual building blocks for research and action', *Water International*, 39 (2), pp. 143–158.

10 Against the trend

Structure and agency in the struggle for public water in Europe

Andreas Bieler

Introduction

Despite the global financial crisis since 2007, neoliberalism has continued to reign supreme. And yet, there is one area which defies the dominant trend: water. From the Cochabamba water war in 2000 to the United Nations declaration of water as a human right in 2010, from the re-municipalization of water in Grenoble in 2000 to the re-municipalization of water in Paris in 2010 and Berlin in 2013, the struggle against water privatization has picked up pace and is a clear success story of resistance to further neoliberal restructuring. Agency against privatization, however, does not take place in a vacuum but is structurally conditioned by capitalist social relations of production.

In this chapter, I will first develop a conceptual discussion of structure and agency in the struggles over water privatization drawing on the philosophy of internal relations (Ollman, 1976). Subsequently, I will assess the structuring conditions of the global economy. Against the background of the global financial crisis, similar to other public services, water has been identified as a potentially secure investment opportunity including also tendencies towards the increasing financialization of water services. Then, the chapter will discuss how we can conceptualise the agency of resistance through a critical engagement with mainstream social movement literature. Drawing on the notion of the social factory developed by Harry Cleaver (2000) as well as feminist Social Reproduction Theory, the chapter will indicate how resistance to water privatization combines struggles at the workplace with struggles within the wider sphere of social reproduction, providing fertile ground for cooperation between trade unions and social movements against capitalist exploitation. The subsequent section will provide empirical examples relating to the struggles over the European Citizens' Initiative 'Water is a Human Right' as well as the resistance to water privatization in Italy and Greece before the conclusion offers some final reflections on these struggles as a basis for societal transformation.

Structure and agency in the struggle over public water

Every approach within the social sciences implicitly or explicitly adopts a conceptualization of the relationship between structure and agency. In this chapter, I will draw on a historical materialist conceptualization of agency and structure in order to acknowledge the historical specificity of capitalism. Mainstream political economy approaches tend to engage in a dualist framing of history across international studies. They identify distinct spheres such as 'agents' and 'structures', 'politics' and 'economics', or 'states' and 'markets' as separate spheres in a relationship of *ontological exteriority* (Morton, 2013: 139–43). In other words, the mainstream political economy literature continuously separates out, in different ways, these binaries. By taking the separate appearance of the state and market as their starting point of analysis, however, they fail to acknowledge the historical specificity of the current capitalist period (Burnham, 1995). Hence, there is the need for a historical materialist moment, in which it is asked why it is that the state and market, the political and the economic appear as separate in the first place (Bieler and Morton, 2018: 3–23). This assertion is attached to a *philosophy of internal relations*, which implies that the character of capital is considered as a social relation in such a way that the internal ties between the relations of production, state-civil society, and conditions of class struggle can be realized. Through this philosophy of internal relations, the dialectical method of historical materialism, therefore, focuses on internally related causes and conditions rather than positing logically independent factors existing side-by-side one another (Ollman, 1976: 48). On the basis of such a philosophy of internal relations, it can then be comprehended that it is the specific ways in which the capitalist social relations of production are organized that makes the state and market appear as separate spheres. Based on wage labor and the private ownership of the means of production, the extraction of surplus labor is not directly politically enforced unlike in feudalism, because those who do not own the means of production are 'free' to sell their labor power (Wood, 1995: 29, 34). Nobody is forced to work for a particular employer. However, without owning one's own means of production, people are indirectly forced to look for paid employment. They are compelled to sell their labor power in order to reproduce themselves. Thus, to understand inequality and exploitation in capitalism, we need to investigate the 'hidden abode of production.' 'We must explore the netherworld of production, outside and beneath the market, where economic necessity compels workers owning only their labor power to seek employment' (Barker, 2013: 44). Unless this historical specificity of our current period is understood, any conceptualization of resistance will fall short of grasping the concrete opportunities of, but also obstacles to, transforming capitalism.

Taking the capitalist social relations of production as an ontological starting point, it can be conceptualized how structure and agency are generated. First, there are the structuring conditions of capitalism, the way production and accumulation of surplus value are set up. At the same time, these social relations of production also engender social class forces as key collective agents. As a result of private property and wage labor, two main classes oppose each other in capitalism: on the one hand capital, the owners of the means of production, and on the other labor, those who are indirectly forced to sell their labor power. It is this focus on the social relations of production as the generator of both structure and agency that facilitates a focus on the internal relations, which allows us to assess the role of class agency within the structuring conditions of capitalism.

Water privatization and the structuring conditions in the global political economy

Due to the way in which capitalist social relations of production are organized around wage labor and the private ownership of the means of production, capitalism is characterized by a set of key structuring conditions (Bieler and Morton, 2018: 38–41). First, because both labor and capital have to reproduce themselves through the market, the resulting compulsion to competitiveness makes capitalism such a dynamic mode of production. At the same time, however, it is also crisis-ridden, the second structuring condition, due to regular 'crises of overaccumulation', when surpluses of both capital and labor can no longer be brought together in a productive way within the capitalist social relations of production (Harvey, 1985: 132). In order to overcome crises, as a third structuring condition, capitalism constantly has to expand outward and to incorporate new, non-capitalist spaces. This outward expansion can be geographical in that new areas are integrated or reintegrated along new lines into capitalism or it can be an inward expansion, in that de-commodified areas are re-commodified for profit making. It is in relation to the third structuring condition that the privatization of public services becomes important. At times when the global economy is in crisis and other investment opportunities have dried up, investing in service provision, ultimately guaranteed by the state, promises sustainable and usually subsidized profits (Fattori, 2013b: 378). As recent reports confirm, international banks and investment institutions have identified water as an excellent, profitable investment opportunity. 'A disturbing trend in the water sector is accelerating worldwide. The new "water barons" – the Wall Street banks and elitist multibillionaires – are buying up water all over the world at unprecedented pace' (Yang, 2012). Slogans such as 'water is the petroleum for the next century' (Goldman Sachs) or 'the water market will soon eclipse oil, agriculture, and precious metals' (Citigroup) drive this new investment fever. Ultimately, 'from the perspective of the financial sector, investing in water is expected to be profitable because of growing demand and constant or diminishing supply which will put upward pressure on the price' (Bayliss, 2014: 301). David Harvey has aptly termed this kind of privatization of resources such as water, formerly held in common, as 'accumulation by dispossession', indicating the ongoing process of primitive accumulation, identified by Marx as the moment when the producer is divorced from the means of production (Harvey, 2003: 145–6).

Private equity funds are playing an increasingly dominant role in the British water sector. 'Of the 10 large water and sewerage companies, four – Anglian, Southern, Thames and Yorkshire – are already owned by private equity or financial groups. Three large companies are still part of groups quoted on the London stock exchange – Severn Trent, South West and United Utilities: of these, Pennon Group, owners of South-West Water, is 46% owned by 6 major financial shareholders' (Hall and Lobina, 2012a: 20). Water distribution as such is not the main concern of these investment funds. They are investing for profit. A key strategy in this respect is leveraging debt. As Aditya Chakrabortty (2014) reported in the Guardian, between 2007 and 2012, there was only one year in which the consortium of shareholders of Thames Water took out less money of the company than it had made in post-tax profits, thereby doubling the company's debt to £7.8bn.

During the 2000s, the shift towards privatization of water services, mentioned previously, started to stall. Private French and English companies withdrew from their international engagements and here in particular from developing countries. In view of insufficient profitability and related criticism by their shareholders, the companies' inability to deliver on their promises and mounting resistance movements, these companies started to look elsewhere for business (Hall and Lobina, 2009: 84–8). The French companies Suez and Veolia 'have instead refocused on engineering work, such as water and wastewater treatment plants and desalination plants, water and sanitation services for industrial companies, and consultancy contracts. Like the companies' other main business, waste management, these are areas of reliably growing demand' (Hall and Lobina, 2012b: 132). English private companies, on the other hand, have increasingly focused on profits via financialization. In this process, investors bought the water companies 'in large part using debt finance, which was then added to the debts of the company (rather than staying with the investors)' (Bayliss, 2017: 387). Moreover, in order to provide investment opportunities satisfying the interest in water by international banks and investment institutions mentioned previously, 'financial innovation has led to the launch of at least four major water-focused exchange traded funds (ETFs) since 2005' (Bayliss, 2014: 298). This securitization of water companies is based on household water bills, which have been repackaged and sold on via offshore jurisdictions, linking water consumers to global financial markets (Bayliss, 2017: 388). In short, the very structural dynamics of capitalism generate significant pressure to privatize public water.

These pressures towards privatization have been further intensified for countries in the European periphery (see also the chapter by van den Berge et al. in this volume). Against the background of the global financial market and Eurozone crises, capital through the agency of the so-called Troika, consisting of European Commission, European Central Bank, and International Monetary Fund, has pushed for further privatization in the so-called bailout agreements.

Imposed austerity policies include: (1) cuts in funding of essential public services; (2) cuts in public sector employment; (3) pushes towards privatization of state assets; and (4) the undermining of industrial relations and trade union rights through enforced cuts in minimum wages and a further liberalization of labor markets (Bieler and Morton, 2018: 233–8). As Panagiotis Sotiris remarks, current European integration is above all 'a class strategy that represents the combined efforts of European capitalist classes to respond to the global economic crisis and to the particular crisis of the European "social model" by means of an offensive neoliberal strategy of capitalist restructuring' (Sotiris, 2017: 172). Across the EU, employers have abused the crisis to cut back workers' postwar gains. Nevertheless, neither do structuring conditions determine agency and prevent resistance nor is the position of capital so dominant that resistance could not make a difference. In the next section, the agency of resistance in struggles over public water is conceptualized.

Agency of resistance in the struggle over water

Within historical materialist scholarship, there is an important critique of structuralist varieties, which overemphasise the structural constraints of capitalism and the power of capital. In a survey of classical Marxist political economy, Harry Cleaver outlines how many of these authors exaggerated the structural constraints on the successful class agency of resistance. 'One must conclude that such theories which accord all power to capital can only be in its interest' (Cleaver, 1979/2000: 46). In short, while not neglecting the structuring conditions of capitalism, through a focus on the internal relations between structure and agency (see the previous discussion), it is necessary to reflect on how to conceptualise agency within a historical materialist framework.

Social movements and civil society more generally have been widely studied by liberal approaches in view of increasing levels of inequality against the background of globalization. In line with Karl Polanyi's (1957) ideas about a double movement, in which a period of laissez-faire is followed by a period of regulation, liberal scholars discuss the possibility of establishing global governance institutions, which can ensure a more just distribution of increasing wealth, resulting from neoliberal restructuring at the global level (e.g. Held and McGrew, 2002: 135–6). There are, however, a number of problems associated with this. First, these scholars understand civil society as some kind of progressive force. It is, however, overlooked that civil society also includes pro-globalization forces such as business associations, which are often a driving force behind global restructuring. Second, liberal analyses overlook the crucial importance of the capitalist social relations of production around the private ownership of the means of production and wage labor. Donatella della Porta and Luisa Parks, for example, focus on changing opportunity structures within the EU, when analysing whether social movements focus on the European or the national level in their campaigns on issues of social justice (della Porta and Parks, 2016, see also Parks, 2015: 22). Of course, institutional structures are important for understanding agency, but understanding why certain structures have been established in the first place and why they might be in the process of changing still requires analysing the underlying social relations of production and how they have conditioned institutional formations. Thus, different organizations have different levels of structural power available, with business organizations in times of transnational production networks being more powerful than national trade unions, for example (Bieler, 2011: 165–70). 'Civil society is not some kind of benign or neutral zone where different elements of society operate and compete freely and on equal terms, regardless of who holds a predominance of power in government' (Buttigieg, 1995: 27).

Hence, in order to analyse the dynamics underpinning water struggles in Europe a historical materialist approach to social movements with an emphasis on social class forces as the main collective actors, and a focus on class struggle as key to understanding economic-political developments is required (Bieler, 2014). By starting the analysis through a focus on the social relations of production, it is understood that the underlying power structures and different levels of resources within capitalist social relations of production engender asymmetries across business, trade union, and social movement groups. In this chapter civil society is understood in a Gramscian sense. Importantly, for Gramsci the form of state consists of 'political society', i.e. the coercive apparatus of the state more narrowly understood including ministries, the police and other state institutions, and 'civil society', made up of political parties, unions, employers' associations, churches, etc. (Gramsci, 1971: 257–63, 271). Civil society is the sphere of hegemonic struggle over the purpose of a particular state form. 'Civil society is simultaneously the terrain of hegemony and of opposition to hegemony' (Buttigieg, 2005: 38). As a Marxist, Gramsci was, of course, aware of the internal relations between the political and the economic and how the underlying production structures resulted in different levels of structural power for agents in civil society.

Marxist class analysis has frequently been criticized for focusing only on struggles at the workplace and the agency of trade unions, thereby overlooking the struggles over gender, race, nationalism, and the environment. Harry Cleaver provides us with the conceptual tools to go beyond such a narrow, one-dimensional analysis. When reflecting on the increasing number of struggles of the late 1960s and 1970s, Cleaver asserts that 'the reproduction of the working class involves not only work in the factory but also work in the home and in the community of homes' (Cleaver, 1979/2000: 70). Hence, the analysis of class struggle has to cover the whole 'social factory', not just the workplace, and includes struggles against the exploitation in the sphere of social reproduction. The notion of 'social factory' is useful, first because it indicates the importance of the social relations of production reflected in 'factory.' By adding 'social,' however, it makes clear that production and reproduction of capital and labor are not only assured within production narrowly understood but extends into the social and natural substratum. In Nancy Fraser's understanding, we need to go beyond the hidden abode of production to capture the full dynamics of capitalist exploitation. She notes 'the institutional separation of "economic production" from "social reproduction", a gendered separation that grounds specifically capitalist forms of male domination, even as it also enables capitalist exploitation of labor power and, through that, its officially sanctioned mode of accumulation' (Fraser, 2014: 67). Hence, when analysing capitalism, we need to analyse all the work, which is required for its ongoing processes of accumulation. This is where social reproduction theorists offer crucial insights. They 'perceive the relation between labor dispensed to produce commodities and labor dispensed to produce people as part of the systemic totality of capitalism' (Bhattacharya, 2017: 2) and are, thus, able to comprehend social struggles in both the spheres of production and reproduction as part of the same contestation over capitalist exploitation and expropriation. For example, 'American workers are fighting to keep their water from being turned off and struggling over their rents, their cost of living, and the state of transportation and education. They fight to keep their neighborhoods safe from racist policy. They fight for access to welfare, health care, and child care. They are organizing against climate change. Some of these struggles are beginning to link up with those in workplaces, once more raising important questions about how social reproduction can act as a site of class recomposition and unity' (Mohandesi and Teitelman, 2017: 65).

Access to water is precisely such an issue, impacting on all the areas of production, reproduction, and ecological destruction. Privatizing water, transforming water into a commodity to make a profit, clearly affects the workplace and generally results in lower wages and deteriorating working conditions for workers employed in the sector. At the same time, however, it also goes beyond and affects every worker as well as the wider community outside work, considering how crucial access to safe drinking water and sanitation is in daily life. Equally, the way we deal with water has a direct impact on the environment and is, thus, crucial for the sustainability of the natural substratum. In short, the struggle against the commodification of water as a tradable, economic good by an alliance of trade unions, social movements, and NGOs is clearly an instance of class struggle within and across the 'social factory.'

Fighting for public water against the background of economic crisis

There are three water struggles in Europe, I want to look at briefly from this perspective on class agency within the wider structuring conditions. Privatization of municipal water services started in Italy in the late 1990s and early 2000s in Tuscany and other locations in central Italy (Bieler, 2015). When prices shot up immediately, local communities started to organize resistance. The Alternative World Water Forum in Firenze in 2003 then opened up space for larger, national-level mobilization. In this encounter, Italian community activists realized that they were facing the same problems as people in the Bolivian city of Cochabamba, for example, prompting reflections about organizing at the national level in opposition to water privatization. Importantly, the emerging alliance resulted in the establishment of the Italian Water Forum in 2006, bringing together social class forces across the whole social factory. Funzione Pubblica, the large public sector union affiliated with the CGIL, as well as the smaller, rank-and-file unions Cobas and USB were part of the alliance. The Comitato Italiano Contratto Mondiale sull'Acqua (CICMA) expressed the concerns of developmental NGOs and the human right to water globally, the environmental movement was represented by the Legambiente and WWF Italia. Water users had mainly organized in local activist groups, which also ensured a territorial presence across the country in addition to the national level, which was essential to mounting a strong campaign at all levels. This alliance first initiated the collection of signatures in 2010, providing them with the institutional right to call for a referendum to abrogate the law by the Berlusconi government in 2009, the so-called 'Decreto Ronchi', enforcing privatization by requiring municipalities to put water contracts out for tender and to establish public-private partnerships with a private participation of at least 40 per cent (Ciervo, 2010: 162). Then, they mobilized widely for the referendum itself, which included two questions. 'The first question cancelled the legal obligation to privatize the management of water services' (Fattori, 2011), i.e. the 2009 law of the Berlusconi government. The second question removed the legal right of private investors to make 7 per cent of their profit from running water services. Together, both questions removed the rationale for private involvement in water distribution. When the referendum took place on 12 and 13 June 2011, the victory of the water movement was overwhelming. For the first time in 16 years, it had been possible to secure the quorum of at least 50 per cent plus one voter participating. In fact, just over 57 per cent of the electorate, more than 26 million Italians, cast their vote. The majorities in relation to the two questions on water were even more impressive. '95.35% yes (4.65% no) on the first question; 95.80% yes (4.20% no) on the second' (Fattori, 2011). The victory could not have been more decisive.

Inspired by successes such as the Italian water referendum or the re-municipalization of water in Paris in 2010, the European Federation of Public Service Unions (EPSU) decided to launch a European Citizens Initiative 'Water and Sanitation are a Human Right' (see the chapter by van den Berge et al. in this volume). Three key objectives were stated at the launch of the ECI in May 2012: '(1) The EU institutions and Member States be obliged to ensure that all inhabitants enjoy the right to water and sanitation; (2) water supply and management of water resources not be subject to 'internal market rules' and that water services are excluded from liberalization; and (3) the EU increases its efforts to achieve universal access to water and sanitation' (see www.right2water.eu). Between May 2012 and September 2013, close to 1.9 million signatures were collected across the European Union (EU) and formally submitted to the Commission. The quotas for the minimum of seven required EU member states were reached in 13 countries. Austria, Belgium, Finland, Germany, Greece, Hungary, Italy, Lithuania, Luxembourg, the Netherlands, Slovakia, Slovenia, and Spain collected the required amount of signatures. Germany stood out as the country with the most signatures. 1,341,061 signatures were collected, of which 1,236,455 were considered valid. Three key reasons can explain the success of the ECI. First, there is this special quality of water, which attracts support across the political spectrum. Already the Italian referendum had shown that a close connection with the Catholic social doctrine, for example, guarantees support from Catholic institutions and people on the centre-right of the political spectrum. Second, the long history of water struggles in Europe, but also beyond, ensured that water was at the top of the agenda and many water activists had already been politicized in previous struggles. Third, while EPSU ensured the tight coordination of the campaign in Brussels and formed an alliance across the social factory including environmental groups and social policy NGOs, similar broad alliances were also established at the various national levels (Bieler, 2017: 305–11).

Finally, while witnessing the hearing of the ECI in the European Parliament through a video link, activists from the Thessaloniki citizens' movement against water privatization decided to hold their own independent referendum about the privatization of water services in their city on 18 May 2014 (see van den Berge et al. in this volume). EPSU, the Italian water movement, as well as others from the European water movement, sent monitors in support. After a large turnout and significant rejection of privatization in this unofficial referendum, with 98 per cent of those who voted opposed to privatization, the pressure on the Greek government not to privatize mounted. Once the Greek Constitutional Court had additionally ruled that the privatization of Athen's water company was unconstitutional, it decided to put a stop to the privatization of water services in both Thessaloniki and Athens (MacroPolis, 2014). The Greek campaign too had been based on a broad alliance across the social factory including trade unions but also a whole range of neighborhood groups as well as social movements (Bieler and Jordan, 2018: 945–6).

In short, when we look at these three cases, we can see how social class forces successfully resisted the privatization and commodification of water. Nevertheless, despite these victories, the outcomes did not fully reflect success. The first referendum question in the Italian case ensured that no further privatizations took place. There was, however, no move towards a re-municipalization of those water services, which had already been privatized, with the exception of Naples. The principle of the EU Stability Pact of balanced budgets was transferred to the level of Italian municipalities. With their financial possibilities constrained, those municipalities, in which water services had already been privatized, found it difficult, if not impossible, to buy back private shares, especially against the background of the Eurozone crisis. The second question of the referendum abrogating the right of private companies to a guaranteed profit of 7 per cent has never really been implemented. In the end, the formula, calculated in exactly the same way, but referred to differently, has been reintroduced at the slightly lower level of 6.4 per cent. The Forum challenged this unsuccessfully in the Administrative Tribunal of Milano in March 2014 (Bieler, 2015). As for the European Citizens' Initiative, the main success was that while the collection of signatures was still taking place, the Commission decided to exclude water from the Concessions Directive, which was being negotiated at the same time. This, in itself, ensured that water services across the EU were excluded from liberalization and, thus, indirectly also from privatization (Fattori, 2013a). Overall, however, the Commission never responded satisfactorily to the ECI, and water as a human right has not been included in EU legislation (Bieler, 2017: 311–15). In Greece, when the third bailout agreement was concluded in July 2015, water was again included in the privatization agenda. The struggle of Greek citizens to protect their water services goes on (European Public Service Union, 2018).

In order to understand the limited success of these struggles despite the clear victories in popular, democratic campaigns, it is important to study the structuring conditions of capitalism, within which these campaigns took place. It was precisely in the second half of 2011, shortly after the referendum, that Italy too increasingly ran into difficulties of refinancing state debt on the financial markets. In turn, the EU and ECB put heavy pressure on Italy towards privatization. In August 2011, Jean-Claude Trichet, the then President of the ECB, and Mario Draghi, who succeeded him in November 2011, urged '"the full liberalization of local public services . . . through large scale privatizations", ignoring the fact that 95.5 per cent of Italian voters had rejected the privatization of local water services in a valid national referendum less than eight weeks earlier' (Erne, 2012: 229). Eventually, in 2012, the Italian government committed itself to austerity measures to receive financial assistance from the EU, but this did not include the liberalization of water services, which had been blocked by a Constitutional Court ruling.

The Greek case is very similar. Here too the Eurozone crisis provided the opportunity for capital to increase the pressure on further water privatization. Initially, the new Syriza government, elected in January 2015, promised an end to austerity for Greek citizens. Over the next six months, however, it was put under heavy pressure from the Troika to live up to austerity measures, agreed with previous Greek governments. In July 2015, when it finally caved in, water was again part of privatization plans (Bieler and Jordan, 2018: 952) and the alliance against privatization fragmented to some extent (see van den Berge, Boelens, and Vos in this volume). It is clearly visible in this case how the agency of resistance runs up against the structuring conditions of capital. The crisis has shifted the balance of power in society to capital, which in turn used the crisis to impose restructuring, which would have been impossible otherwise. This does not imply that resistance is impossible. The struggle in Greece continues with an open-ended outcome, but it demonstrates that agency on its own does not guarantee success. At the European level, water is safe from privatization for now, as it was, for example, excluded from the Comprehensive Economic and Trade Agreement (CETA) (with Canada). Nonetheless, the limited response by the Commission indicates that here too the interests of (transnational) capital and the ongoing economic crisis imply that the privatization of water services can return onto the agenda at any moment.

Water and the potential for transformation

In this chapter, I have developed a position on structure and agency through the philosophy of internal relations, which allows us to study concrete moments of class struggle over water privatization within the historically specific structuring conditions of capitalism. Resistance is possible, indeed has had some significant success, but ultimately remains contested, as the crisis of overaccumulation within the European political economy persists.

Some observers criticise the strategy of declaring water to be a human right as individualistic and, thus, counterproductive to a potential collective response to privatization (e.g. Bakker, 2010: 13, 158–9). Jamie Linton, however, understanding the human right to water as a relation, highlights the transformative potential of water struggles on the basis of an expanded definition of the human right to water. First, he defines the right to water as a rule of governance. 'Treating water "as a social and cultural good, and not primarily as an economic good" implies that the basic questions of how water is allocated and managed should be decided by democratic processes rather than market principles' (Linton, 2012: 50). Second, he defines the human right to water as a rule of social equity. Hence, 'the right to water entails the right of the collective to a share of the value generated in the hydrosocial production process' (Linton, 2012: 57). Importantly, both dimensions can be detected in ongoing water struggles.

First, the Italian referendum campaign was characterized by the underlying discourse around water as a commons, understood as 'elements that we maintain or reproduce together, according to rules established by the community: an area to be rescued from the decision-making of the post-democratic elite and which needs to be self-governed through forms of participative democracy' (Fattori, 2011). It directly challenged the capitalist focus on commodifying ever more areas and submitting them to the profit logic of the market, implying a move towards a new economic model. Social equity and the right of everyone to share in the value of water production is noticeable. This focus was combined with a new, participatory form of democracy in the running of water services, reflecting Linton's reference to democratic processes in water management. It is this new understanding of democracy and a new way of how to run the economy and, importantly, of how these two dimensions are closely and internally related, which brings with it a transformative dimension. The focus on the commons was picked up in Thessaloniki, where members of the water movement in the group K136 developed the idea that if every household connected to the city's water service bought a non-transferable share in the state-owned water company of Thessaloniki, the public could own the water and sanitation company through a system of neighborhood cooperatives coming together through a single overall cooperative (Bieler and Jordan, 2018: 950–1). Working on an alternative model of how to run the city's water services, the group emphasized the importance of a new form of democracy. 'The model is based on direct democracy, meaning that decisions are taken at open assemblies and are based on the principles of self-management and one person, one vote' (Steinfort, 2014). To develop alternatives of this type and perhaps even extend them beyond water into areas of transport, health services, or education has proved to be extremely difficult. Nonetheless, these efforts indicate a potentially transformative way beyond capitalist social relations of production. Within the structuring conditions of capitalism, water struggles, thus, provide a glimpse of a possible alternative for the future.

References

Burnham, P. 1995State and market in international political economy: Towards a Marxian alternative *Studies in Marxism* 2135–159

Bhattacharya, T. (2017) 'Introduction: Mapping social reproduction theory', in Bhattacharya, T. (ed.) *Social reproduction theory: Remapping class, recentering oppression.* London: Pluto Press, pp. 1–20.

Bakker, K. (2010) *Privatizing water: Governance failure and the world's urban water crisis*. Ithaca, NY: Cornell University Press.

Barker, C. (2013) 'Class struggle and social movements', in Barker, C., Cox, L., Krinsky, J. and Nilsen, A.G. (eds.) *Marxism and social movements*. Leiden and Boston, MA: Brill, pp. 41–61.

Bayliss, K. (2014) 'The financialization of water', *Review of Radical Political Economics*, 46 (3), pp. 292–307.

Bayliss, K. (2017) 'Material cultures of water financialisation in England and Wales', *New Political Economy*, 22 (4), pp. 383–397.

Bieler, A. (2011) 'Labour, new social movements, and the resistance to neo-liberal restructuring in Europe', *New Political Economy*, 16 (2), pp. 163–183.

Bieler, A. (2014) 'Transnational labour solidarity in (the) crisis', *Global Labour Journal*, 5 (2), pp. 114–133.

Bieler, A. (2015) '"Sic vos non vobis" (for you, but not yours): The struggle for public water in Italy', *Monthly Review*, 67 (5), pp. 35–50.

Bieler, A. (2017) 'Fighting for public water: The first successful European Citizens' Initiative "Water and sanitation are a human right"', *Interface: A Journal for and about Social Movements*, 9 (1), pp. 300–326. Available at: www.interfacejournal.net/wordpress/wp-content/uploads/2017/07/Interface-9-1-Bieler.pdf

Bieler, A. and Jordan, J. (2018) 'Commodification and "the commons": The politics of privatising public water in Greece and Portugal during the Eurozone Crisis', *European Journal of International Relations*, 24 (4), pp. 934–957.

Bieler, A. and Morton, A.D. (2018) *Global capitalism, global war, global crisis*. Cambridge: Cambridge University Press.

Buttigieg, J.A. (1995) 'Gramsci on civil society', *Boundary 2*, 22 (3), pp. 1–32.

Buttigieg, J.A. (2005) 'The contemporary discourse on civil society: A Gramscian critique', *Boundary 2*, 32 (1), pp. 33–52.

Chakrabortty, A. (2014) 'Thames water: The drip, drip, drip of discontent', *The Guardian*, June 15. Available at: www.theguardian.com/commentisfree/2014/jun/15/thames-water-discontent-privatisation

Ciervo, M. (2010) *Geopolitica dell'Acqua* (new edition). Rome, Italy: Carocci editore.

Cleaver, H. (2000) *Reading capital politically (second edition)*. Leeds: Anti/Theses.

della Porta, D. and Parks, L. (2018) 'Social movements, the European crisis, and EU political opportunities', *Comparative European Politics*, 16 (1), pp. 85–102. Available at: doi:10.1057/s41295-016-0074-6

Erne, R. (2012) 'European industrial relations after the crisis: A postscript', in Smismans, S. (ed.) *The European Union and industrial relations: New procedures, new context.* Manchester: Manchester University Press, pp. 225–235.

European Public Service Union (2018) 'The fight for public water goes on in Greece' February 26. Available at: www.epsu.org/article/fight-public-water-goes-greece

Fattori, T. (2011) 'Fluid democracy: The Italian Water Revolution', *transform!*, October 27. Available at: http://transform-network.net/journal/issue-092011/news/detail/Journal/fluid-democracy-the-italian-water-revolution.html

Fattori, T. (2013a) 'The European Citizens' Initiative on Water and "Austeritarian" Post-Democracy', *transform!*, December 13. Available at http://transform-network.net/journal/issue-132013/news/detail/Journal/the-european-citizens-initiative-on-water-and-austeritarian-post-democracy.html

Fattori, T. (2013b) 'From the Water Commons Movement to the commonification of the public realm', *The South Atlantic Quarterly*, 112 (2), pp. 377–387.

Fraser, N. (2014) 'Behind Marx's hidden abode: For an expanded conception of capitalism', *New Left Review II*, 86, pp. 55–72.

Gramsci, A. (1971) *Selections from the prison notebooks*, ed. and trans. Q. Hoare and G. Nowell Smith. London: Lawrence and Wishart.

Hall, D. and Lobina, E. (2009) 'Water privatization', in Arestis, P. and Sawyer, M. (eds) *Critical essays on the privatization experience.* London: Palgrave, pp. 75–120.

Hall, D. and Lobina, E. (2012a) *Water companies and trends in Europe 2012.* London: Public Services International Research Unit.

Hall, D. and Lobina, E. (2012b) 'The birth, growth, and decline of multinational water companies', in Katko, T., Juuti, P.S., and Schwartz, K. (eds) *Water services management and governance.* London: IWA Publishing, pp. 123–132.

Harvey, D. (1985) 'The geopolitics of capitalism', in Gregory, D. and Urry, J. (eds) *Social relations and spatial structures.* London: Macmillan, pp. 128–163.

Harvey, D. (2003) *The new imperialism.* Oxford: Oxford University Press.

Held, D. and McGrew, A. (2002) *Globalization/anti-globalization.* Cambridge: Polity.

Linton, J. (2012) '"The human right to what?" Water, rights, humans, and the relation of things', in Sultana, F. and Loftus, A. (eds) *The right to water: Politics, governance, and social struggles.* London/New York: Routledge, pp. 45–60.

MacroPolis (2014) 'Greece shelves water privatisation plans, leaving gap in revenue targets', July 2. Available at: www.macropolis.gr/?i=portal.en.economy.1331

Mohandesi, S. and Teitelman, E. (2017) 'Without reserves', in Bhattacharya, T. (ed.) *Social reproduction theory: Remapping class, recentering oppression.* London: Pluto Press, pp. 37–67.

Morton, A.D. (2013) 'The limits of sociological Marxism?', *Historical Materialism*, 21 (1), pp. 129–158.

Ollman, B. (1976) *Alienation: Marx's conception of man in capitalist society* (2nd ed.). Cambridge: Cambridge University Press.

Parks, L. (2015) *Social movement campaigns on EU policy: In the corridors and in the streets.* London: Palgrave.

Polanyi, K. (1957) *The great transformation: The political and economic origins of our time.* Boston, MA: Beacon Press.

Sotiris, P. (2017) 'The authoritarian and disciplinary mechanism of reduced sovereignty in the EU: The case of Greece', in Tansel, C.B. (ed) *States of discipline: Authoritarian neoliberalism and the contested reproduction of capitalist order*. London: Rowman & Littlefield International, pp. 171–187.

Steinfort, L. (2014) 'Thessaloniki, Greece: Struggling against water privatisation in times of crisis', June 3. Available at: www.tni.org/article/thessaloniki-greece-struggling-against-water-privatisation-times-crisis

Wood, E.M. (1995) *Democracy against capitalism: Renewing historical materialism*. Cambridge: Cambridge University Press.

Yang, J.-S. (2012) 'The new "water barons": Wall Street mega-banks are buying up the world's water' December 12. Available at: www.marketoracle.co.uk/article38167.html

11 Remunicipalization and the human right to water

A signifier half full?

David A. McDonald

Introduction

Debates about the human right to water are hotly contested and nowhere is this truer than on the question of water privatization. Supporters of private sector involvement in water services argue that it advances the human right to water, arguing that private companies are more efficient, more accountable and better equipped than the public sector to extend affordable water services to all and therefore ensure equal access (Aquafed 2017, Salman & McInerney-Lankford 2004).

Private water companies have been quick to apply these human rights arguments. When the United Nations was first considering water as a human right in 2010, French multinational Veolia was amongst the first to declare its position: "Water is a human right. This is the strong belief of Veolia and the core of a water operator job: bring water to those who need it most" (Veolia 2010, 1). Suez was equally resolute: "We want to take part in the promotion and implementation of the right to water and sanitation. We're able to offer a full range of solutions in response to all issues faced by both developed and developing countries".[1] After years of bad press, the human right to water offered private water companies and their supporters an opportunity to give privatization a positive spin.

Opponents of privatization, meanwhile, argue that privatization undermines the human right to water by making it unaffordable to the poor, by cutting people's services off for non-payment, by violating environmental regulations, and by failing to provide adequate levels of service provision (Food & Water Watch 2011, Shiva 2016, Bakker 2009). The debate could hardly be more polarized.

Formal international law on the subject offers little to resolve this impasse. In fact, it is unequivocally neutral on the topic, with the UN Office of the High Commission on Human Rights (OHCHR 2010, 35) arguing that "international human rights law does not prescribe whether water services should be delivered by public or private providers or by a combination of the two", only that states are required to "guarantee equal access to affordable, sufficient, safe and acceptable water. . . . if water services are operated or controlled by third parties".

In other words, the international legal system offers no conceptual or practical assistance when it comes to the question of public versus private and the human right to water. This agnostic approach is further evidenced in the silence this topic receives in official UN statements on the topic. UN Water, for example, makes no mention of public or private provision in its position on water as a human right, and the same can be said of the UN's Special Rapporteur on the topic.[2] Many large water NGOs are equally ambivalent. WaterAid (2011, 4, 15) has had a "rights-based approach" at the center of its Global Strategy since 2005, but argue that "Human rights law does not take sides on the public versus private debate"; what is required is "regulatory systems to monitor these impacts, regardless of whether services are provided by a public or private entity".

But even if the UN did take a position on public versus private water there would be little to compel member states to adhere to it in practice. UN agencies cannot impose rules on member states about how to achieve the human right to water and would be unlikely to apply sanctions in a world where many other human rights abuses go unaddressed. Little wonder, then, that Sultana and Loftus (2015, 98) worry about the right to water "becoming an empty signifier used by both political progressives and conservatives who are brought together within a shallow postpolitical consensus that does little to effect real change in water governance."

Could the trend towards remunicipalization of water help tip the balance in this regard? Cities around the world are taking water services back under public management and ownership after years of private sector control, with at least 267 cases of water remunicipalization in 37 countries over the past 15 years, affecting more than 100 million people (Kishimoto and Petitjean 2017). The pace of remunicipalization appears to be growing (Lobina 2017), and there is an expanding international movement in favour of publicly-managed water.

This trend prompted the Chair of *Eau de Paris* (which remunicipalized in 2010) to claim that "a counter-attack is underway and is spreading throughout the world…giving rise to a new generation of public companies that are beginning to play a stronger role in the water sector" (Blauel 2015, 2). The majority of remunicipalization has thus far occurred in two countries – France and the US – but it is a truly global phenomenon, including cities as diverse as Accra (Ghana), Almaty (Kazakhstan), Antalya (Turkey), Budapest (Hungary), Buenos Aires (Argentina), Conakry (Guinea), Dar es Salaam (Tanzania), Kuala Lumpur (Malaysia) and La Paz (Bolivia). Half of all cases have occurred since 2010, suggesting an acceleration of interest (Lobina 2015).

There is no guarantee that public water provision will advance the human right to water, of course. Some 90% of the world's water services are already in public hands but there are still 700 million people without adequate access to drinking water and 2.4 billion without adequate access to sanitation (World Health Organization/United Nations Children's Fund 2015). Austerity and other forms of public service cutbacks are to blame for much of this lack of delivery, but corruption and poor governance are partly responsible as well. In other words, there is nothing inherently equitable or rights-oriented about state-owned and state-managed water.

The same can be said for remunicipalization. As we shall see in the following sections, some forms of remunicipalization are autocratic in nature, serving to reinforce or worsen inequalities. Others are neoliberal in character, using public ownership as a way to commercialize water services and advance a larger expansion of marketized social relations in ways that could not be accomplished via privatization. There are more progressive forms of remunicipalization, seeking to decommodify water and expand access and enhance equity, but these too are not without their problems when it comes to the human right to water.

The remainder of this chapter examines these different types of remunicipalization and asks: 1) how (and if) the human right to water has been articulated within them, and 2) whether or not they can effectively advance the cause of the human right to water and promote a "public" approach to achieving them. Drawing on a previous typological framework by the author (McDonald 2018), we look at five different remunicipalization models, illustrated by concrete examples of how human rights discourses are applied.

There is no "perfect" fit between remunicipalization and the human right to water or any "one size fits all" model. My aim here is simply to highlight the synergies and tensions between remunicipalization and the human right to water movements and to help shed light on the different ways in which they have been implemented. In doing so, I hope to contribute to what Sultana and Loftus (2015, 101) see as a "key challenge" in the application of the human right to water, by helping "to fill this empty signifier with real political content".

In this regard, I think the glass is half full when it comes to remunicipalization. Real progress has been in some cases, but remunicipalization is not inherently progressive; it is contested and contradictory, and should be seen as part of a larger struggle for more equitable, sustainable and accountable public water provision.

Remunicipalization and the human right to water

Not all remunicipalizations happen by choice. There are many instances in which policymakers would prefer to have private service provision but are forced to remunicipalize because of an insufficient number of (credible) private sector bidders for a contract. One example is that of Hamilton, Canada, where, in 2004, efforts to renew a private contract failed because there were no companies willing to bid on what were deemed to be overly restrictive contract conditions, obliging the municipality to bring water services back in-house, against the ideological inclinations of its elected officials (González-Gómez, García-Rubio, & González-Martínez 2014, Ohemeng & Grant 2008). In other cases, private firms are unwilling to bid on what they see to be unprofitable contracts. There are also examples of private companies ending contracts early, compelling governments to remunicipalize. Such was the case in Buenos Aires, Argentina, in 2000, when a private consortium headed by Suez ended its contract with the city prematurely due to macro-economic instability in the country and frustration with its lack of profits (Azpiazu & Castro 2012, de Gouvello, Lentini, & Brenner 2012).

Having said that, the majority of remunicipalizations appear to be planned and deliberate. Many are driven by dissatisfaction with private sector service performance, including concerns with rising costs to consumers, worsening service quality, non-achievement of infrastructure promises, public mistrust of private companies, anti-trust activities on the part of large private utilities, and corruption (Estache & Grifell-Tatjé 2010; FWW 2010; Hall, Lobina, & Motte 2005; Hall, Lobina, & Corral 2010; Hall, Lobina, & Terhorst 2013, Le Strat 2014, Lobina et al. 2014, Pérard 2009, Valdovinos 2012, Ruiz-Villaverde & García-Rubio 2017, Warner 2010, Wollman et al. 2010).

In other cases, municipalities may be satisfied with the quality of private service but choose to remunicipalize because of the high costs of monitoring and regulating private contracts. This is true of large, long-term concessions as well as small, short-term contracts, all of which require sophisticated and expensive regulatory management (if they are to be done well). Far from reducing the costs and complications of service delivery, many municipalities are discovering that contracting out requires costly teams of lawyers and bureaucrats, reducing or even reversing potential efficiency gains (Bel et al. 2010, Le Strat 2014, Nickson & Vargas 2002, Wu & Ching 2013).

But this disappointment with the costs of privatization conceals a much more diverse set of remunicipalization ideologies. Saving money and improving services might be central to most remunicipalization initiatives, but this seemingly common agenda hides a much more complex set of philosophical starting points, underscored by the ways in which they adopt (or ignore) the human right to water.

Autocratic state capitalism

The first type of remunicipalization we explore is that of autocratic state capitalism, so-called to denote instances where the reversal of privatization is undertaken by relatively undemocratic, but market-oriented, governments as part of a larger shift back towards state control of strategic sectors in a capitalist economy. In these cases, the remunicipalization of water is driven as much by political and social objectives as economic ones, ranging from attempts to enhance national sovereignty to regulating ethnic minorities.

In some respects, this is a very old storyline, with the control of water being at the heart of many different forms of "despotic" regimes over the centuries, with "unaccountable, unregulated and, above all, undemocratic" forms of state water governance intended to enhance control by a ruling elite (Strang 2016, 294). What makes this particular form of water autocracy different is its grounding in market ideology and its use of commercialized management techniques, with publicly-owned water intended to enhance market growth at the same time as it extends socio-political control. As such, this form of remunicipalization is not necessarily anti-private in its orientation. Rather, it should be seen as a strategic reversal of privatization, under certain conditions, with the aim of achieving targeted social goals while expanding market-like operational characteristics such as full cost recovery and financially-driven performance indicators to enhance other market functions in the economy.

Not surprisingly, the human right to water does not feature strongly in the (admittedly limited) public statements and research papers on this type of remunicipalization. The cities of Almaty and Astana in Kazakhstan illustrate this point. After a brief period of contracted privatization in the early 2000s, both jurisdictions converted back to public management after a Presidential Decree declared water to be a strategic resource that must remain public, part of a larger push to nationalize key assets and centralize power in the country.[3] The human right to water does not feature in either of these initiatives and does not feature prominently in the water sector in the country, but a vague reference the Water Code of Kazakhstan that a "water consumer. . .has the right to use water resources to meet his own needs".[4] The fact that the Government of Kazakhstan abstained from the UN vote in 2010 on the Universal Declaration of the Human Right to Water (Martín 2017, 112) is another telling feature, as are widespread accusations of other human rights violations (Human Rights Watch 2018).

More subtle, but equally problematic, is the case of Hungary. Despite a robust academic debate on the human right to water in the country (Szabó & Greksza 2013; Szilágyi 2013), and a Constitutional commitment to "ensuring access to healthy [. . .] drinking water" (WaterLex 2014, 63), the right to water did not feature strongly in the decision to remunicipalize water in Budapest in 2012. Nor has the conservative nationalist government ruled out future privatizations. According to Horváth (2016, 193), the motivation for remunicipalization was driven largely by "economic factors" and a determination to lower tariffs for populist reasons (see also Bencze and Mindak 2016).

There are other examples of autocratic forms of remunicipalization which may offer additional insights eg. Antalya (Turkey), Bamako (Mali), Conakry (Guinea) (Lobina 2015; World Bank 2006), but a lack of empirical data precludes any deeper analysis at this point in time. It should be noted that these are a small minority of remunicipalization cases to date, but the growth of state capitalisms in general, and the potential for privatization reversals in China in particular, may see these figures rise, with little effort to advance the human right to water.

Market managerialism

The second type of remunicipalization is that of market managerialism. This model is also aimed at promoting markets and advancing capital accumulation, but in these cases, the rationale for putting water services back into state hands largely revolves around enhancing efficiency. Grounded in a neo-Keynesian reading of context-specific market failures (eg. insufficient competition, lack of regulatory capacity on the part of the state), private-sector service delivery is seen to be less efficacious than state delivery in these cases, creating a drag on the economy as a whole (Stiglitz 1991). Remunicipalization is seen as a necessary (if temporary) measure to reduce operating costs and ensure sufficient investment in services to expand local production and consumption. It is also seen as an opportunity to create an entrepreneurial water operator: one with cost recovery, internal competition and marketized forms of managerial incentives at its heart, part of a longer-term shift towards new public management, resulting in a "broadening and blurring of the "frontier" between the public and private sectors", combined with a "shift in value priorities away from universalism, equity, security and resilience towards efficiency and individualism" (Pollitt 2003, 474). Remunicipalized water services driven by this logic can be characterized as quasi-commercial entities, focusing on market-based performance indicators. They may be public in name but can serve to deepen, not weaken, the commercialization of water, while at the same time attacking the perceived failures of Keynesian-era welfarism (Clarke et al. 2007).

When the human right to water appears in the rationale for these forms of remunicipalization it takes on a decidedly liberal tone, one that categorizes rights as individualized and reciprocal, offering "consumers" an opportunity to compete in a market of rights while downplaying the structural inequities that prevent individuals and groups from participating equally (on this liberal discourse see Bakker 2007). The World Bank's (2016, np) human rights rhetoric captures this liberal perspective perfectly:

> Does the World Bank Group recognize the Human Rights to Water and Sanitation? Yes. . . . Does the human right to water and sanitation mean that water and sanitation services should be free for everyone? No, rather it implies that water and sanitation must be affordable for all and nobody to be deprived of access because of an inability to pay. As such, the human rights framework does not provide for a right to free water. . . . We believe a utility company should deliver services across all consumers in their area and whilst doing so they must be accountable, efficient, and environmentally, socially and financially sustainable.

An example of such marketized forms of remunicipalization and human rights is that of Dar es Salaam in Tanzania. Starting in the 1990s, the Tanzanian government, with a loan from the World Bank, "designed and implemented a new administrative water rights and fee payment system with the aim of improving basin-level water management and cost-recovery for government water-resource management services" (Van Koppen et al. 2007, 143). The Bank also advised the government to experiment with public-private partnerships in water, but after a disastrous experience in Dar es Salaam in 2003, they reversed their policy recommendations, promoting instead the creation of a new corporatized public water operator in 2005. The Dar es Salaam Water and Sewerage Corporation has since managed to extend coverage and improve some aspects of service delivery – "proving that public water services can be managed well by the state, and can outperform the private sector in many ways" – but the new public entity has become much more market-oriented than it was before privatization, enforcing cost recovery on the poor and "failing to meet its obligations in the lowest income areas of the city" (Pigeon 2012a, 41). Rights legislation, meanwhile, focuses largely on "responsibility to pay" (Triche 2012, 7).

These commercialized public water operators are common in Africa (where the World Bank and other neoliberal donor agencies remain influential) as well as the United States (Warner 2016), although experience in the US shows little evidence of any kind of rights-based discourse when it comes to reversing privatization. National surveys with US city managers consistently show "cost savings" to be the primary motive for moving back to public ownership (Warner and Hefetz 2012). Whether this is merely bureaucratic pragmatism – as opposed to an ideological preference for state delivery – is unclear, but an emphasis on corporatized forms of public water in the US is strong, and may expand elsewhere as privatization failures force pro-market governments to seek commercialized in-house alternatives with little in the way of a human rights discourse to shape the way this unfolds [the intransigent situation with contaminated water in Flint, Michigan, is illustrative of this point (Butler, Scammell, and Benson 2016)].

Social democracy

The third type of remunicipalization can be broadly defined as social democratic. This is the most common (and most celebrated) of the remunicipalization categories outlined in this chapter and represents the bulk of water remunicipalization in Western Europe and Latin America (Kishimoto and Petitjean 2017, 6). It also has some of the most extensive languages around the right to water.

In general, these cases of remunicipalization entail more robust state intervention than marketized forms, with the aim of promoting social and economic justice. Cost-reflexive pricing and other market-management tools are often still employed in these models, but they are typically combined with a commitment to challenging commodification and advancing values of water beyond its marginal price. There also tends to be a strong commitment to equity via cross-subsidization and ensuring better access to water services across a range of social, spatial and economic divides.

These broad social democratic principles are captured in the following excerpt from the "Declaration for the Public Management of Water" signed by the Mayors of Madrid, Barcelona and eight other Spanish cities in November 2016 (Cities for Public Water 2016, np, *emphasis added;* note in particular the explicit link made between human rights and public water):

1 We believe that water and its associated ecosystems are a common good that cannot be appropriated for the benefit of private interest. All of nature's good and resources form part of the natural

patrimony of the planet and are indispensable for the sustainment of life, which obligates us to preserve and protect them. We therefore defend that they be managed with criteria of solidarity, environmental sustainability, mutual cooperation, collective access, equity and democratic control, without contemplating profit.

2 *We fully assume that the provision of water for human consumption and sanitation is a human right which, in accordance with the doctrinal body of the United Nations, is indispensable for a dignified life and a precondition for the realization of other fundamental rights.* As a result, supplies must be provided on a universal basis, guaranteeing a vital minimum for all people and ensuring that any supply cuts for economic or social reasons are prohibited.

3 We understand that the management of supply and sanitation services must, in addition to being necessarily public, promote new forms of social control that guarantee transparency, information, accountability and effective citizen participation. *As a result we are committed to a management model in which the public entity responsible for these services accounts for their activities and actions, both to public authorities and citizens, and that, both in day-to-day management, in planning and in decision making process, instruments of citizen participation are established in the services, linked to the urban water cycle, promoting the necessary consensus.*

4 As a result, we reject the privatization of the integral urban water cycle services and we support the re-municipalization processes that are being carried out in numerous cities and towns to recover the public, democratic and transparent management of water to guarantee a rendering of accounts to the citizens.

Eau de Paris has also placed human rights at the center of its social democratic mandate. When the city of Paris took back control of its water system from Suez and Veolia in 2010, new regulations were written to "affirm the principle of the basic human right to water. That is why the municipality and its management company refuse to cut off water to any occupied housing, even if illegally occupied housing, until so ordered by Court decision. For social and sanitation reasons, water supply for everyone must be a full and unalienable right" (Mairie de Paris 2013, 24). *Eau de Paris* has also worked to advance gender equity in the workplace, improve the protection of upstream water resources through partnerships with farmers, and develop public-public partnerships with service providers in Morocco, Mauritania, and Cambodia. They have promoted water conservation (despite its impacts on lowering water revenues), while at the same time creating a water solidarity fund to assist low-income households [with some 44,000 homes receiving benefits worth 500,000 euros in 2011 (Petitjean 2015, 67, Pigeon 2012b, 36)].

The European Citizens Initiative (ECI) on the Right2water (www.right2water.eu) is another example of how the right to water fits with the social democratic remunicipalization movement is. Spearheaded by organized labor along with a broad coalition of NGOs and community associations, the ECI asked the European Commission to "propose legislation implementing the human right to water and sanitation as recognized by the United Nations, and promote the provision of water and sanitation as essential public service for all". They collected close to 1.9 million signatures in Europe and demanded that "EU institutions and Member States be obliged to ensure that all inhabitants enjoy the right to water and sanitation; water supply and management of water resources not be subject to 'internal market rules' and that water services are excluded from liberalization; and that the EU increases its efforts to achieve universal access to water and sanitation". Although not a direct rejection of privatization, the ECI has nevertheless "changed the public discourse on water in Europe [and . . .] arguments about the importance of keeping water in public hands would no longer be laughed at or belittled" (Bieler 2017, 313).

Nor are these principles of public water and human rights confined to Europe. After renationalizing its water services in the early 2000s, Uruguay became the first country in the world to enshrine the right to water in its Constitution (Moshman 2005). Jakarta is another example, with the Supreme Court of Indonesia having ordered the government to restore public water services to residents after finding private companies "failed to protect" their right to water. The regional government has been ordered to revoke the private contracts and "hand responsibility for public water supply services back to a public water utility" (Harsono 2017, np, see also Kooy & Bakker 2008, Kooy 2014).

But as positive as these changes have been it cannot be forgotten that social democratic forms of remunicipalization are not explicitly anti-market in their objectives, and continue to operate within the constraints of a broader capitalist economy. Social democratic rights can improve access to and affordability of water, but they cannot reverse the broader commodification process nor fundamentally alter larger socio-economic inequalities (Esping-Anderson 1990).

Uruguay, for example, remains captured by corporatist politics, with social movements having been "subsumed under the left government's political project, which prioritizes international trade and continues the corporatist tradition of the Uruguayan state, thus limiting the scope of reform and restricting participation by civil society and the water sector trade union" (Terhorst, Olibera, and Dwinell 2013, 60–1). In Jakarta, critics worry that a "full remunicipalization is unlikely to happen until the contract expires in 2023" (Marwa and Tobing 2018, np), and the new owner of the private water company is a well-connected Indonesian conglomerate, making it unclear if the transition to a remunicipalized public water entity will ever take place, despite the Supreme Court's position on human rights (Hodge and Rayda 2017).

Such social democratic tensions are not unique to the water sector, of course, but they tend to be under-acknowledged in the remunicipalization movement, with the inherent limitations of human rights legislation within a market economy bound to generate rifts in the struggle against privatization in the future.

Anti-capitalism

There are also remunicipalization movements and organizations that are driven by explicitly anti-capitalist sentiments. These groups tend to share many of the same goals as their social democratic counterparts – including the human right to water – but they reject the possibility of a reconciliation between water justice and capitalism, pointing to the many ways in which market economies colonize our broader lifeworlds.

These anti-capitalist voices are not uncommon in the water remunicipalization movement but are seldom in the ascendency, with anti-capitalist actors having yet to realize an actual remunicipalization victory. This lack of success may not be surprising in a world of neoliberal hegemony, but is exacerbated by the fact that anti-capitalist positions on water services tend to be highly fragmented, struggling to find a unified vision of what a "socialist" water project might look like, driven as much by a rejection of old-style communism as they are by a denunciation of the market. A growing commitment to grassroots voices, transparent decision-making, and smaller-scale infrastructure development provides some cohesiveness, but as with anti-capitalist political movements more broadly, there is as much that pulls them apart as binds them together when it comes to (re)building public water and (re)defining what is meant by human rights (Tormey 2012).

Bolivia is an emblematic case in point. The Water Wars of the early 2000s brought together an eclectic coalition of organizations in the struggle to oust Bechtel from Cochabamba – some of which were grounded in radical anti-market politics (Olivera and Lewis 2004) – but these anti-capitalist voices struggled to make a difference on the ground when it came to restructuring the new public water operator. Even the election of President Evo Morales in 2005 – a self-proclaimed socialist who proposed the Resolution on the Human Right to Water at the UN General Assembly in 2010 – has not managed to create a socialist vision of the human right to water in practice. Although the Bolivian state is "progressively realizing the [human right to water] by global standards of access and investment levels, the broader criteria for [this right], including citizen participation and democratic decision making, remain largely unfulfilled" (Baer 2015, 353).

Sadly, post-privatization water reforms in Cochabamba remain stymied by a gridlock of elite politics, neoliberal logic, bureaucratized decision-making, and social democratic forces, with little in the way of transformational change. Much of this stasis can be blamed on conservative forces in the city, but many also blame a national government which has "not fully broken with neoliberal policies" despite its socialist rhetoric: "[M]ass mobilizations have been disempowered [and] the relations between movements and governments have been full of tensions and unmet expectations. . . . [P]rogressive discourse on the environment and water has typically been accompanied by unilateral decisions, a lack of debate, and the criminalization and stigmatization of critical voices" (Terhorst, Olibera, & Dwinell 2013, 56, 63; see also Cameron & Hershberg

2010; Spronk & Webber 2007). As a result, the human right to water remains central to anti-capitalist discourses in the city but has yet to be realized in non-marketized ways.

Autonomism

Finally, there are advocates of remunicipalization that are leery of both capitalist and socialist forms of change (Mazzoni & Cicognani 2013, Spronk, Crespo, & Olivera 2012, Bélanger, Spronk, & Murray 2016). There are overlaps here with other types of remunicipalization (including, once again, a push for the human right to water) but this category distinguishes itself with its emphasis on community-driven water service solutions grounded in a local socio-ecological context with little or no direct state involvement. In fact, these are not remunicipalization movements, *per se*, because they are generally opposed to centralized and bureaucratized forms of state water delivery (regardless of its ideological orientation) and are often opposed to a take-over by the state (González-Gómez et al. 2014, Heller et al. 2007, Driessen 2008, Laurie & Crespo 2007, Gorostiza, March, & Sauri 2013, Marston 2015).

There are also philosophical differences in the meaning of the human right to water. Many autonomous groups push for "customary rights", determined locally and organically through traditional water practice, and actively resist incorporation into a public legal system (Gupta, Ahlers, & Ahmed 2010, Ostrom 2008, Trawick 2003). Localized notions of water norms are seen to be ontologically and practically incompatible with a universal declaration, particularly one articulated and monitored by the state.

Once again, Bolivia is emblematic. During the Water Wars in Cochabamba, peri-urban farming communities were a major part of the anti-privatization coalition. But these groups wanted to reclaim their usufruct rights to water, a form of collective management based on social agreements negotiated and renegotiated over time known as *usos y costumbres* (uses and customs) (Boelens, Bustamante, & Perreault 2010, Marston 2015). After water was remunicipalized, these groups continued to push for these customary rights and continue to resist incorporation into a municipal water system (Terhorst, Olibera, and Dwinell 2013).

Similar dynamics have emerged in Mexico. The Mexican government enshrined the right to water in its Constitution in 2012, noting that "Everyone has the right to access clean, good quality water in sufficient quantities and at affordable rates for their personal and domestic use, together with sanitation services for its disposal" but this was followed by the assertion that "The state will guarantee this right" (Frenk 2018, np). The law allows for the "participation of citizens for the achievement of such purposes", but community groups have since mobilized to define their own demands and mechanisms of participation and decision making. A Citizens' Initiative for the General Water Law was created, composed of community organizations, indigenous groups, NGOs, academics, and water professionals, demanding a much wider range of rights, including "guaranteed access to water both for human communities and for all other species and ecosystems. . .minimum quantity of water necessary to meet their day-to-day needs, regardless of their purchasing power, age, sex and place of residence" as well as a principle of participation that "affirms that it is at the local level, in the management of springs, streams, wells and water intakes in each community or neighborhood, where it is possible to reinforce people's knowledge, their forms of organization and the mechanisms they use to plan, run and supervise water management"; in other words a radical decentralization of knowledge, power and the concept of a human right (Frenk 2018, np). As a result, many rural Mexicans remain skeptical of what the human right to water means in practice. As Frenk (2018, np) notes, municipal government "management procedures usually rule out participation by local residents and instead foster clientelism and corruption". In these cases, remunicipalization can be a dirty word.

Conclusion

We have seen in this chapter that there is nothing inherently pro-human rights about the remunicipalization of water. Nor are there common narratives across the remunicipalization spectrum, suggesting that the human right to water may indeed be the "empty signifier" that Sultana and Loftus (2015, 98) fear.

And yet, the concept of a right to water continues to mobilize and animate progressive water movements around the world, serving as a rallying point for anti-privatization initiatives in particular. It is critical that these movements continue to challenge the notion that profit-seeking private enterprises can meet the needs of citizens in equitable and sustainable ways. As such, the right to water remains a useful conceptual and practical tool for organizing against privatization.

But the anti-privatization movement cannot rest here. The bigger challenge is figuring out what type(s) of public provision can help us move towards the actualization of the right to water. Some public models can be easily dismissed (such as the autocratic examples cited previously) but the other types of remunicipalization outlined in this paper present a more significant philosophical and strategic challenge.

Commercialized forms of public water present a particularly difficult dilemma for water rights activists. Although public in name, many have been fundamentally altered by the ideological and institutional practices of neoliberalism, and it can be difficult to see how they might be changed to be more equity-oriented in the short to medium term. It took decades to create these corporatized entities; they will not be changed again overnight. Nevertheless, heavily commercialized public water operators must be exposed for their market-like practices, with more progressive narratives on the human right to water being used to highlight their problems.

Equally challenging will be what to do with social democratic forms of remunicipalization. Here we see significant progress on a wide range of indicators, with many positive human rights gains. But can we expect these public services to sustain their practices in a world of deepening commodification and growing commercial demands on water supplies? Anti-capitalist and autonomous activists argue this is impossible, but also recognize that we are a long way from realizing more radical anti-market and anti-state movements in the near future. This is perhaps why radical water activists continue to collaborate with social democratic organizations to achieve concrete gains in the short run. How long these coalitions can last is difficult to say, but they will be tested as the contradictions and inequities of market dynamics continue to play themselves out.

So, what is to be done? In the short term, it is important to impress upon multilateral agencies such as the UN, the World Bank, and mainstream NGOs that public water operators can do a better job than private companies at realizing the human right to water. . ..*if* they are adequately funded, *if* they are transparent in their processes and *if* they commit themselves to democratic engagement with citizens about what the right to water means in practice. Weak, corrupt and highly commercialized public water agencies are not human rights advocates, and reclaimed public water operators cannot change themselves overnight. But if global agencies responsible for funding water initiatives were to invest as much time and money into improving and democratizing public water as they have in promoting private sector investment there could be a significant generational change.

One optimist trend is the growth of public-public partnerships (PUPs), with progressive public water operators cooperating with each other to advance the human right to water (Boag & McDonald 2010, Silvestre, Marques, & Gomes 2018, Baer 2017). Tensions exist within these partnerships, but frank debates about the possibilities and limitations of different forms of public water are essential to realizing the right to water in the future.

Notes

1 See www.suez.com/en/Who-we-are/A-committed-group/We-support-the-right-to-water-and-sanitation
2 See www.unwater.org/water-facts/human-rights and www.ohchr.org/EN/Issues/WaterAndSanitation/SRWater/Pages/SRWaterIndex.aspx respectively.
3 www.remunicipalisation.org/#case_Almaty
4 http://adilet.zan.kz/eng/docs/K030000481_

References

Aquafed (2017) The human rights to water and sanitation: overcoming the challenges to implementation. Available at: www.aquafed.org/Public/Files/publication/submission_to_world_water_council_rtws_research_project_aquafed_final_8f78e2961b.pdf

Azpiazu, D. and Castro, J.E. (2012) "Aguas públicas: Buenos Aires in muddled waters", in Pigeon, M., McDonald, D.A., Hoedeman, O., and Kishimoto, S. (eds) *Remunicipalisation: putting water back into public hands*. Amsterdam: Transnational Institute, pp. 58–73.

Baer, M. (2015) From water wars to water rights: implementing the human right to water in Bolivia, *Journal of Human Rights*, 14 (3), pp. 353–376.

Baer, M. (ed) (2017) Democratizing water? Public-public partnerships in the Global South in Munoz-Garcia, F. (ed) *World scientific reference on natural resources and environmental policy in the era of global change* (Vol. 2: The social ecology of the Anthropocene: Continuity and change in global environmental politics). New York: World Scientific Publishing, pp. 323–344.

Bakker, K. (2007) The "commons" versus the "commodity": alter-globalization, anti-privatization, and the human right to water in the Global South, *Antipode*, 39 (3), pp. 430–455.

7Bakker, K. (2009) The "commons" versus the "commodity": alter-globalization, anti-privatization, and the human right to water in the Global South, *Antipode*, 39 (3), pp. 38–63.

Bel, G., Fageda, X., and Warner, M.E. (2010) "Is private production of public services cheaper than public production? A meta-regression analysis of solid waste and water services", *Journal of Policy Analysis and Management*, 29 (3), pp. 553–577.

Bencze, T. and Mindak, E. (2016) Experiences of Budapest Waterworks with state, municipal ownership structures and with the involvement of private funding: case study of Budapest Waterworks, *Water Practice and Technology*, 11 (1), pp. 58–65.

Bélanger, M.D., Spronk, S., and Murray, A. (2016) "Work of the ants: labour and community reinventing public water in Colombia", in McDonald, D.A. (ed) *Making public in a privatized world: the struggle for essential services*. London: Zed Books, pp. 26–42.

Bieler, A. (2017) Fighting for public water: the first successful European Citizens' Initiative, "Water and sanitation are a human right". *Interface: A Journal for and about Social Movements*, 9 (1), pp. 300–326.

Blauel, C. (2015) Foreword, in Kishimoto, S., Lobina, E., and Petitjean, O. (eds) *Eau publique, eau d'avenir*. Amsterdam: Transnational Institute, pp. 1–4.

Boag, G. and McDonald, D.A. (2010) "Critical review of public-public partnerships in water services", *Water Alternatives*, 3 (1), pp. 1–25.

Boelens, R., Bustamante, R. and Perreault, T. (2010) "Networking strategies and struggles for water control: from water wars to mobilizations for day-to-day water rights defence", in Boelens, R., Getches, D., and Guevara-Gil, A. (eds) *Out of the mainstream: Water rights, politics and identity*. New York: Routledge, pp. 281–305.

Butler, L.J., Scammell, M.K., and Benson, E.B. (2016). The Flint, Michigan, water crisis: a case study in regulatory failure and environmental injustice, *Environmental Justice*, 9 (4), pp. 93–97.

Cameron, M.A. and Hershberg, E. (eds) (2010) *Latin America's left turns: politics, policies, and trajectories of change*. Boulder, CO: Lynne Rienner Publishers, pp. 98–127.

Cities for Public Water (2016) Consolidation of the public water movement in Spain: Outcome of the first meeting Cities for Public Water. Available at: www.tni.org/en/article/consolidation-of-the-public-water-movement-in-spain

Clarke, J., Newman, J., Smith, N., Vidler, E., and Westmarland, L. (2007) *Creating citizen-consumers: changing publics and changing public services*. Thousand Oaks, CA: Pine Forge Press.

de Gouvello, B., Lentini, EJ., and Brenner, F. (2012) Changing paradigms in water and sanitation services in Argentina: towards a sustainable model?, *Water International*, 37 (2), pp. 91–106.

Driessen, T. (2008) Collective management strategies and elite resistance in Cochabamba, Bolivia, *Development*, 51 (1), pp. 89–95.

Esping-Anderson, G. (1990) *The three worlds of welfare capitalism*. Cambridge: Polity Press.

Estache, A. and Grifell-Tatjé, E. (2010) *Assessing the impact of Mali's water privatization across stakeholders* (ECARES Working Paper 2010–2037).

Food and Water Watch (2010) "The public works: how the remunicipalization of water services saves money". Available at: www.scribd.com/document/53983183/The-Public-Works-How-the-Remunicipalization-of-Water-Services-Saves-Money

Frenk, G.A. (2018) "Flowing movement": building alternative water governance in Mexico. Available at http://longreads.tni.org/state-of-power-2018/flowing-movement-building-alternative-water-governance-in-mexico

González-Gómez, F., García-Rubio, M.A., and González-Martínez, J. (2014) Beyond the public–private controversy in urban water management in Spain. *Utilities Policy*, 31, pp. 1–9.

Gorostiza, S., March, H., and Sauri, D. (2013). Servicing customers in revolutionary times: the experience of the collectivized Barcelona Water Company during the Spanish Civil War, *Antipode*, 45 (4), pp. 908–925.

Gupta, J., Ahlers, R., and Ahmed, L. (2010) The human right to water: moving towards consensus in a fragmented world, *Review of European Community & International Environmental Law*, 19 (3), pp. 294–305.

Hall, D., Lobina, E., and Corral, V. (2010). *Replacing failed private water contracts* (PSIRU report). London: University of Greenwich. Available at: http://gala.gre.ac.uk/2761/1/PSIRU_Report_9823_-_2010-01-W-Jakarta.pdf

Hall, D., Lobina, E., and Motte, R.D.L. (2005) Public resistance to privatisation in water and energy, *Development in Practice*, 15 (3–4), pp. 286–301.

Hall, D., Lobina, E., and Terhorst, P. (2013) Re-municipalisation in the early twenty-first century: water in France and energy in Germany, *International Review of Applied Economics*, 27 (2), pp. 193–214.

Harsono, A. (2017) Indonesia's Supreme Court upholds water rights: court rules Jakarta's water privatization failed the poor, October 12. Available at: www.hrw.org/news/2017/10/12/indonesias-supreme-court-upholds-water-rights

Heller, L., Moraes, L.R.S., Borja, P.C., de Melo, C.H., and Sacco, D. (2007) *Successful experiences in municipal public water and sanitation services from Brazil*. Sao Paolo, Brazil: National Association of Municipal Water and Sanitation Services. Available at: www.tni.org/en/publication/successful-experiences-in-municipal-public-water-and-sanitation-services-from-brazil

Hodge, A. and Rayda, N. (2017). Fears new Jakarta rulers will go to water reversing privatization, *The Australian*, October 13. Available at: www.theaustralian.com.au/news/world/fears-new-jakarta-rulers-will-go-to-water-reversing-privatisation/news-story/2ce84bb1cddccb419b6f11699d1c7c71

Horváth, T.M. (2016) From municipalisation to centralism: changes to local public service delivery in Hungary, in Wollmann, H., Koprić, I., and Marćou, G. (eds), *Public and social services in Europe: from public and municipal to private sector provision*. London: Palgrave, pp. 185–200.

Human Rights Watch (2018) "World Report 2018: Kazakhstan, events of 2017". Available at: www.hrw.org/world-report/2018/country-chapters/kazakhstan

Kishimoto, S. and Petitjean, P. (2017) *Reclaiming public services: how cities and citizens are turning back privatization*. Amsterdam: Transnational Institute.

Kooy, M. (2014) "Developing informality: the production of Jakarta's urban waterscape", *Water Alternatives*, 7 (1), pp. 35–53.

Kooy, M. and Bakker, K. (2008) Splintered networks: the colonial and contemporary waters of Jakarta, *Geoforum*, 39 (6), pp. 1843–1858.

Laurie, N. and Crespo, C. (2007) Deconstructing the best case scenario: lessons from water politics in La Paz–El Alto, Bolivia, *Geoforum*, 38 (5), pp. 841–854.

Le Strat, A. (2014) Discussion – The remunicipalization of Paris's water supply service: a successful reform, *Water Policy*, 16 (1), pp. 197–204.

Lobina, E. (2015) Calling for progressive water policies, in Kishimoto, S., Lobina, E., and Petitjean, O. (eds) *Our public experience: the global experience with remunicipalisation*. Amsterdam: Transnational Institute, pp. 6–18.

Lobina, E. (2017) Water remunicipalisation: between pendulum swings and paradigm advocacy, in Bell, S., Allen, A., Hofmann, P., and Teh, T.H. (eds) *Urban water trajectories*. Cham, Switzerland: Springer International Publishing, pp. 149–161.

Mairie de Paris (2013) *Water in Paris: a public service*. Available at: www.eaudeparis.fr/uploads/tx_edpevents/Brochure_institutionnelle_ENG_2013.pdf

Marston, A.J. (2015) Autonomy in a post-neoliberal era: community water governance in Cochabamba, Bolivia. *Geoforum*, 64, pp. 246–256.

Martín, M.Á.P. (2017) *Security and human right to water in Central Asia*. New York: Springer.

Marwa and Tobing, D.H. (2018) Jakarta's plan to get more public power in water sector might not work well, *The Conversation*, 8 February. Available at: http://theconversation.com/jakartas-plan-to-get-more-public-power-in-water-sector-might-not-work-well-89320

Mazzoni, D. and Cicognani, E. (2013) Water as a commons: an exploratory study on the motives for collective action among Italian water movement activists, *Journal of Community & Applied Social Psychology*, 23 (4), pp. 314–330.

McDonald, D.A. (2018) Remunicipalization: the future of water services?, *Geoforum*, 91, pp. 47–56.

Moshman, R. (2005) "The constitutional right to water in Uruguay", *Sustainable Development Law and Policy*, 5 (1), p. 15.

Nickson, A. and Vargas, C. (2002) The limitations of water regulation: the failure of the Cochabamba concession in Bolivia. Bull, *Latin American Research*, 21 (1), pp. 99–120.

Ohemeng, F.K. and Grant, J.K. (2008) "When markets fail to deliver: an examination of the privatization and de-privatization of water and wastewater services delivery in Hamilton, Canada", *Canadian Public Administration*, 51 (3), pp. 475–499.

Olivera, O. and Lewis, T. (2004) *Cochabamba!: water war in Bolivia*. Boston, MA: South End Press.

Ostrom, E. (2008) Institutions and the environment, *Economic Affairs*, 28 (3), pp. 24–31.

Petitjean, O. (2015) Taking stock of remunicipalisation in Paris, in Kishimoto, S., Lobina, E., and Petitjean, O. (eds) *Our public water future: the global experience with remunicipalisation*. Amsterdam: Transnational Institute, pp. 56–74.

Pérard, E. (2009) "Water supply: Public or private? An approach based on cost of funds, transaction costs, e ciency and political costs", *Policy and Society*, 27 (3), pp. 193–219.

Pigeon, M. (2012a) From fiasco to DAWASCO: remunicipalising water systems in Dar es Salaam, Tanzania, in Pigeon, M., McDonald, D.A., Hoedeman, O., and Kishimoto, S. (eds) *Remunicipalisation: putting water back into public hands*. Amsterdam: Transnational Institute, pp. 40–57.

Pigeon, M. (2012b) Une Eau Publique pour Paris: symbolism and success in the heartland of private water, in Pigeon, M., McDonald, D.A., Hoedeman, O., and Kishimoto, S.

(eds) *Remunicipalisation: putting water back into public hands.* Amsterdam: Transnational Institute, pp. 24–39.

Pollitt, C. (2003) *The essential public manager.* Buckingham: Open University Press/McGraw Hill.

Ruiz-Villaverde, A. and García-Rubio, M.A. (2017) "Public participation in European water management: from theory to practice", *Water Resource Management*, 31 (8), pp. 2479–2495.

Salman, S.M.A. and McInerney-Lankford, Siobhan. (2004) *The human right to water: legal and policy dimensions (English)* (Law, justice, and development series). Washington, DC: World Bank.

Shiva, V. (2016) *Water wars: privatization, pollution, and profit.* Berkeley, CA: North Atlantic Books.

Silvestre, H.C., Marques, R.C., and Gomes, R.C. (2018) Joined-up government of utilities: a meta-review on a public–public partnership and inter-municipal cooperation in the water and wastewater industries, *Public Management Review*, 20 (4), pp. 607–631.

Spronk, S. and Webber, J.R. (2007) Struggles against accumulation by dispossession in Bolivia, *Latin American Perspectives*, 34 (2), pp. 31–47.

Spronk, S., Crespo, C., and Olivera, M. (2012). Struggles for water justice in Latin America: public and "social-public" alternatives, in McDonald, D.A. and Ruiters, G. (eds), *Alternatives to privatization: public options for essential services in the global South.* New York: Routledge, pp. 421–452.

Stiglitz, J. (1991) The economic role of the state: efficiency and effectiveness, in Hardiman, T.P. and Mulreany, M. (eds) *Efficiency and effectiveness in the public domain: the economic role of the state.* Dublin: Institute of Public Administration, pp. 37–59.

Strang, V. (2016) Infrastructural relations: water, political power, and the rise of a new "despotic regime", *Water Alternatives*, 9 (2), pp. 292–318.

Sultana, F. and Loftus, A. (2015) The human right to water: critiques and condition of possibility, *Wiley Interdisciplinary Reviews: Water*, 2 (2), pp. 97–105.

Szabó, M. and Greksza, V. (eds) (2013) *Right to water and the protection of fundamental rights in Hungary.* Pécs: University of Pécs Press.

Szilágyi, E.J. (2013) Affordability of drinking-water and the new Hungarian regulation concerning water utility supplies, in Szabó, M. and Greksza, V. (eds) *Right to water and the protection of fundamental rights in Hungary.* Pécs: University of Pécs Press, pp. 68–83.

Terhorst, P., Olivera, M., and Dwinell, A. (2013) Social movements, left governments, and the limits of water sector reform in Latin America's left turn, *Latin American Perspectives*, 40 (4), pp. 55–69.

Tormey, S. (2012) *Anti-capitalism.* Hoboken, NJ: John Wiley & Sons, Ltd.

Trawick, P. (2003) Against the privatization of water: an indigenous model for improving existing laws and successfully governing the commons, *World Development*, 31 (6), pp. 977–996.

Triche, T. (2012) *A case study of public-private and public-public partnerships in water supply and sewerage services in Dar es Salaam* (World Bank Water Paper No 69032, April). Washington, DC: World Bank.

United Nations Office of the High Commission for Human Rights (2010) The right to water. Available at: www.ohchr.org/EN/Issues/WaterAndSanitation/SRWater/Pages/SRWaterIndex.aspx

Valdovinos, J. (2012) "The remunicipalization of Parisian water services: new challenges for local authorities and policy implications", *Water International*, 37 (2), pp. 107–120.

Van Koppen, B., Sokile, C.S., Lankford, B., Hatibu, N., Mahoo, H., and Yanda, P.Z. (2007) Water rights and water fees in rural Tanzania, in Molle, F. and Berkoff, J. (eds) *Irrigation water pricing: the gap between theory and practice*. New York: Colombia International Water Management Institute, pp. 143–163.

Veolia (2010) Right to water: from concept to reality. Available at: www.ohchr.org/ Documents/Issues/Water/ContributionsPSP/Veolia.pdf

Warner, M. (2016) Pragmatic publics in the heartland of capitalism: local services in the United States, in McDonald, D.A. (ed), *Making public in a privatized world: the struggle for essential services*. LondonZed Books, pp. 175–196.

Warner, M. and Hefetz, A. (2012) In-sourcing and outsourcing: the dynamics of privatization among us municipalities2002–2007, *Journal of the American Planning Association*, 78 (3), pp. 313–327.

WaterAid (2011) *Rights-based approaches to increasing access to water and sanitation* (WaterAid Discussion paper). Available at: https://washmatters.wateraid.org/publications/rights-based-approaches-to-increasing-access-to-water-and-sanitation

WaterLex (2014) *National human rights institutions and water governance: compilation of good practices*. Geneva: WaterLex.

Wollmann, H., Baldersheim, H., Citroni, G., Marcou, G., and McEldowney, J. (2010) "From public service to commodity: the demunicipalization (or remunicipalization?) of energy provision in Germany, Italy, France, the UK, and Norway", in Wollmann, H. and Marcou, G. (eds) *The provision of public services in Europe, between state, local government and market*. Cheltenham/Northampton: Edward Elgar, pp. 168–190.

World Bank (2006) *Protecting and improving the global commons: 15 years of the World Bank Group Global Environment Facility Program*. Washington, DC: World Bank.

World Bank (2016) FAQ: World Bank Group support for water and sanitation solutions, February 3. Available at: www.worldbank.org/en/topic/water/brief/working-with-public-private-sectors-to-increase-water-sanitation-access

World Health Organization/United Nations Children's Fund (2015) *Progress on sanitation and drinking water: 2015 update and MDG assessment*. Geneva: Joint Monitoring Program.

Wu, X. and Ching, L. (2013) The French model and water challenges in developing countries: evidence from Jakarta and Manila, *Policy and Society*, 32 (2), pp. 103–112.

12 Citizen mobilization for water

The case of Thessaloniki, Greece

Jerry van den Berge, Rutgerd Boelens and Jeroen Vos

Introduction

The European Citizens' Initiative (ECI) 'Right2Water' collected nearly 1.9 million signatures between September 2012 and September 2013 to put the Human Right to Water and Sanitation on the European political agenda. The ECI is a tool established by the Lisbon Treaty as a means to bring the European Union closer to its citizens. It gives people an opportunity to bring an issue to the European political agenda if they manage to collect over one million signatures in one-year, from at least seven EU countries, with a specific minimum for each country (European Commission, 2011a). A broad European civil society coalition called for the European Commission to develop legislation that would ensure the realization of the human right to water and sanitation across member states. Further, it called for the Commission to contribute to universal access to water and sanitation for all and a halt to the liberalization and privatization of water services. Right2Water united a hugely diverse group of over 250 organizations and was supported by thousands of people that campaigned all over Europe. It gave new momentum to social movements that were active on water issues and extended the focus of some of the social movements that did not pay attention to water until that moment. Right2Water became the first successful ECI.

The ECI passed the threshold of the minimum number of signatures in 14 of 27 countries in which it campaigned. One of the countries where Right2Water was successful was Greece. In the end, 36,000 signatures were collected against a threshold of 16,500. It was a remarkable result because in the first instance, from September 2012 until March 2013, no signature had been collected in Greece. Trade union contacts in Greece noted that Greek citizens were disconsolate at the time because of the austerity measures imposed upon the country and the crisis situation. This changed when, during the second round for the Greek bailout, privatization of the water companies of Athens and Thessaloniki became one of the top priorities in the measures proposed by the Troika of the IMF, the European Union and the European Central Bank. Water privatization was pushed especially vis-à-vis Greece and Portugal (Hall and Lobina, 2012). Right2Water had by then gained a lot of media attention and support of many people in Europe as the conflict with the European Commission proposal for a concession directive was exposed. Liberalization of water services was part of this directive and was explained as a 'privatization through the back door' by Right2Water campaigners. This co-incidence appeared to be a significant influence on both campaigners in Greece as well as for campaigners at the European level. Signature collection skyrocketed in two months and this did not go unnoticed in Greece. In Thessaloniki people organized among different lines to protest against the Troika's measure of privatizing water and a clear link could be established between EU policy, water privatization and Greece. Seeing that in western European countries people protested against the privatization of water services gave the Greek activists the boost that they needed to mobilize people against the Troika-imposed austerity measures that had almost knocked them down.

The Right2Water movement campaigned against the intention of the European Commission to further privatize drinking water utilities in Europe. Privatizing water provision services was, and still is, encouraged by the International Monetary Fund and the World Bank, which makes public-to-private takeovers a condition of lending. As a result, the early 1990s saw a rush of cities and countries around the world signing over their nations' water resources to private companies. It is argued by industry and investors that putting water in private hands translates into improvements in efficiency and service quality, and that services will be better managed (see, e.g. Bakker, 2010; van den Berge et al., 2018). Privatizing also provides governments with an opportunity to gain revenues by selling off water services, and for companies to generate a profit. But with profit being the main objective, the idea of water as a human right arguably becomes a secondary and often forgotten concern. In many cases, problems with water privatization often began to occur soon after the initial wave of enthusiasm – from lack of infrastructure investment to environmental neglect (Lawson, 2015). Privatization of water provision services can also lead to increased prices for consumers, lack of service, slack repairs and unequal access to the water service (Boelens, Perreault and Vos, 2018). Although those who support neoliberal policies advocate a retreat of the state from the market, in reality, governments at national as well as European levels are closely involved in the processes of privatization: "Without the various state levels paving the way and imposing conditions that guarantee privatization and then secure profitable operation afterwards, this accumulation by dispossession could not possibly take place" (Swyngedouw, 2005: 89).

Water in Europe is subject to both European and national law: it is therefore a shared responsibility between the European Union and Member States, making it a suitable issue for an ECI. The initiative aimed to shift the focus of the European Commission from their market-orientation to a rights-based and people-oriented approach to water policy. Right2Water joined in the ongoing struggle for water justice that, in divergent ways, was framed and organized by many civil society groups, and it took a stance against profit-driven water companies with the slogan "Water is a public good; not a commodity!" (van den Berge et al., 2018).

In Greece water services have been in public hands, as was established in water law in 1980. The water companies of Athens (EYDAP) and Thessaloniki (EYATH), that together provide more than half of the population with water services, are stock listed with a majority of shares owned by the state of respectively 61% and 74%. They are both well-functioning and economically-profitable water companies. The other half of the water services are provided by municipal companies called DEYA.

Prior to the ECI, the human right to water had not been a subject of debate in Greek policies and the provision and management of water services was left to local levels. In 2010 the Greek government abstained in the vote at the United Nations General Assembly (UNGA) that recognized water and sanitation as a human right, without explanation (UNGA, 2010). The Greek constitution states that water sources, as vital to life, should be under public control. However, various governments have had plans for privatizing water services since the nineties following the European neoliberal approach to the economy, including the belief that the single market functions best if governments facilitate the operation of market forces and borders are opened without barriers for businesses across the whole of Europe. A privatization agenda has been part of the European agenda of opening markets in public services since the early nineties (Hall & Lobina, 2012; van den Berge et al., 2018). Liberalization and privatization of local and national public services have been proposed by the European Commission since the Maastricht treaty in 1991.

The Greek crisis and water service privatization in Thessaloniki and Athens

Following the financial crisis, indebted countries within the EU, and especially Greece, Ireland and Portugal, were bailed out by the Troika in exchange for the imposed restructuring of their national economies, including labor market deregulation, cutting of public sector employment and privatization of public companies (Lapavitsas et al., 2012). Italy too came under pressure in the second half of 2011, when Jean-Claude Trichet, then President of the ECB, and Mario Draghi, who succeeded him in November 2011, urged "the full liberalization of local public services (. . .) through large scale privatizations" (Bieler, 2015: 9).

The push for privatization in Greece during the Eurozone crisis took place in the framework of austerity measures as a supposed answer to the economic crisis (Zacune, 2013). Between 2008 and 2015, Greece's Gross Domestic Product fell by 29.6 percent (Organisation for Economic Co-operation and Development, 2017). The Greek economy had the largest contraction of any advanced economy since the 1950s (Financial Times, 2015). The severity of the economic downturn in Greece created a more explicit push towards the privatization of public water and sanitation services (Bieler and Jordan, 2017).

In 2010 the European Commission, together with IMF and ECB ordered the Greek government to sell a number of public assets as part of the (first) bailout, that is, extending financial support to the country to avoid bankruptcy

(European Commission, 2010). The Greek government planned to reduce its shares in the water utilities EYDAP and EYATH to 51%. This privatization of water companies met with huge resistance. In Thessaloniki, people from the trade union and employees of EYATH came together with local NGOs to discuss possible actions to prevent the privatization of the company. The trade union would take the lead in mobilizing people in Thessaloniki; the first step in what became a movement later. The most common argument against water privatization concerned tariff increases, which occurred in the vast majority of cases, making safe water inaccessible for many (Lawson, 2015).

The privatization of Greek state-owned enterprises was proposed to ensure a reduction in "subsidies, other transfers or state guarantees", while also leading to "an increase in efficiency of the companies and an extension in the competitiveness of the economy as a whole" (European Commission, 2011b: 33). Interestingly, the Athens Water Supply and Sewerage Company (EYDAP) and the water company of Thessaloniki (EYATH), both earmarked for privatization, have historically been profitable. As Yiorgos Archontopoulos from SOSteTo-Nero stated: "the forced privatization is providing an opportunity to foreign investors to take over our profitable and good functioning public utilities for a real bargain. In five years time, the investment of 40 million Euros will be gained back through the annual and consistent profits of 8 million of EYATh" (Archontopoulos pers. comm., 15 February 2018).

Greece's 'debt crisis' intensified existing social antagonisms, and consequently exacerbated conflicts. While elites and the mass media were trying to drag the population into a collective guilt trip over 'Greek people living beyond their means', a national social engineering operation was set in motion, dispossessing and excluding the bulk of the population. Most importantly, state assets and infrastructure were sold to the highest bidder. The wages, pensions, labor rights and welfare arrangements of the popular classes were therefore slashed overnight (Karyotis, 2017).

Although Greece was not the only country affected by the crisis, it had nevertheless undergone one of the lengthiest and most intense programs of austerity in Europe after 2010. The framing of the crisis as "a national and moral problem" (Mylonas, 2014: 305) that can be blamed on an "overgenerous welfare state" and on "the laziness of people" (Pentaraki, 2013: 701) contributed to boosting authoritarian, nationalistic and xenophobic ideas and practices. The privatization of state assets has always been an integral feature of Greece's international bailouts. In three rounds of bailouts between 2010 and 2015 Greece faltered on promises to sell vital parts of its infrastructure – ports, airports, marinas and waterworks – in exchange for billions of euros in loans. Details of exactly what Greece was required to privatize emerged in August 2015, with the leaking of the 'Memorandum of Understanding for a three-year ESM program' prepared by the Troika (European Commission, 2015). The leaked document listed 23 state assets, ranging from airports to service utilities, and presented precise steps and timelines for privatization (Hellenic Republic Asset Development Fund [HRADF], 2015). This list included two large public water companies: EYDAP and EYATH. Under threat of being forced out of the Eurozone, Athens agreed to transfer 'valuable assets to an independent fund (HRADF), with the aim of raising €50 bn. The privatization fund was the issue that almost forced a Grexit (Rankin and Smith, 2015). Resistance to privatization grew, forcing the new Syriza government to indicate that they would not cross this limit.

Resistance and the rise of social movements

The first initiatives in Greece towards politically decisive resistance over water service provision came from the country's second largest city, Thessaloniki. Here the preliminary steps towards privatization in 2007 were slowed down in part through the resistance by the water workers' union, which staged a four-day hunger strike during the city's international trade fair. The first tenders were eventually announced in 2009 and again the union – which, unlike most unions in Greece, had determinedly maintained its autonomy from all political parties – responded with a 12-day occupation of the company's main building. The water workers union's alliance with activists in Thessaloniki proved to be a foundation on which the growing anti-privatization campaign could build (Wainwright, 2013). The EYATH water workers union saw that privatization plans would not only cut jobs but would also make profits disappear. Being small in terms of citizen numbers, to change the political landscape they searched for coalitions with municipal actors and Civil Society Organizations (CSOs). The workers invested heavily in time to discuss the ideal water provision for citizens. Thus, when, after 2010, the proposal for privatization was brought back to the table (again) they had no difficulty in mobilizing people to combat this plan.

In the spring of 2011, Thessaloniki's 'indignados' movement orchestrated a mass mobilization with well over half a million people taking to the streets to defy austerity and demand democracy. In particular, the protest cohered around opposition to the push for privatization. The water workers of EYATH and a group of citizens set up the 'Initiative 136' (K136) to get water into the hands of the people. K136 opposed the privatization of the Water and Sanitation Company EYATH in Thessaloniki and proposed its social management through local-level cooperatives. The idea of K136 was that every water user would buy a non-transferable share. If one divides the estimated value of EYATH by the number of the users, the result would be the symbolic number of 136 euros, an idea born during the discussions that took place in Thessaloniki, in the so-called White Tower assemblies of the 'squares movement' (known as the 'Indignados'). The different movements decided to join forces in this common endeavor with four purposes:

- The acquisition of 40% of the shares and the management of EYATH by the citizens
- Social control of the city's water provision

- Democratic operation of the company
- To strive for a non-profit based character of the company, in line with social policies and environmental protection[1]

On 9 February 2012, the Greek Parliament ratified the second 'rescue' memorandum and turned it into State Law. With this law, all the privatization procedures that had been debated during eight months were accelerated, causing a severe deterioration in the standard of living of the Greek people and the selling off of national assets. Under the auspices of the International Monetary Fund and the European Commission, the Greek government put on public tender, among many others assets, the Water and Sanitation Company of Thessaloniki (EYATH), even though it had been an efficient and profitable company that offered quality water services for low prices, without ever reporting any financial loss (EYATH, 2016).

In 2013 'SOSteToNero' (SOS for water) was set up by the same water workers that made the link between Right2Water and the fight against privatization of the water companies in Greece. They believed it was necessary to combat privatization both at the local as well as the EU level and gave the momentum to the ECI in Greece. They emphasized the importance of this struggle being part of a wider European movement (Bieler and Jordan, 2017). The EYATH workers' union was central to the establishment of this city-based umbrella social movement organization that brought together water, environmental and social activists and organizations. At the same time, 'Save Greek Water' was set up in Athens. Their argument was that water is essential to life and that selling the water company to foreign investors meant selling life. Even a partial sale could not be accepted. The president of the Thessaloniki water workers' union, George Archontopoulos, feared that private investors "will be given management control as a present". Therefore "whether it is 49% or 51%, we oppose further privatization of the company" (Kishimoto and Hoedeman, 2015). A water movement developing across Attica, the region of Athens, converged on this same target; the imminent threat of privatization. They insisted on the autonomy of their movement from all political parties which, in part, was the response to a history in which independent civil society had been suffocated by the two main political parties.

All Greek organizations united in SOSteToNero and Save Greek Water joined the Right2water campaign to show citizens how EU policy threatened local public control over water. At the same time, the ECI campaigners could point at Greece to indicate that European policies were undermining the confidence in and support for the European project. The success of developing a united front against the privatization of public water and sanitation in Thessaloniki meant that SOSteToNero had become an overarching social movement that coordinated the activities of all its constituent organizations (Bieler and Jordan, 2017). Individual activists of the Athens and Thessaloniki water company unions had developed novel political practices in an attempt to address the disconnection between their organization and wider civil society. They changed the framework in claiming their right to water, transforming their role from powerless citizens in a bankrupt country to self-confident people defining the way that the commons and their right should be governed (Kaika, 2017). This was part of a process of emancipating union activists "from the hierarchies, dependencies and pervasive forms of domination associated with a state operating through clientelism" (Wainwright, 2014: 27). Equally important, union activists shifted their political practice towards so-called 'social movement unionism' (Kretsos, 2011) – a form of unionism that attempts to integrate workers, trade unions and the wider labor movement into broader coalitions for social and economic justice (see also, Perreault, Boelens and Vos, 2018). Such a strategy aligned with the building of a multi-actor and multi-scalar water movement in Greece: The water movement did not emerge alone. It was intrinsically related to the wider social, ideological and political movement that emerged during the 'Greek crisis.' For example, food distribution networks were set up, which inspired the K136 campaign (Calvário, Velegrakis and Kaika, 2016).

Right2Water came at a crucial moment. It took, however, time and endurance to show people in Greece that they were not alone in their struggle and to generate signatures for the European Citizens' Initiative Right2Water. In turn, Right2Water campaigners supported the Greek organizations in organizing a referendum against the privatization of the Thessaloniki water company EYATH. After seeing the massive support for Right2Water in the European Parliament (at the hearing on 17 February 2014), activists in Thessaloniki felt strong enough to call for a referendum. This confidence was not visible in Athens. On 18 May 2014, coinciding with the first round of the local elections, the citizens of Thessaloniki organized a popular, non-binding, referendum to give the chance for the people to express their opinion on the sale of the municipally-owned water company EYATH. The referendum was inspired by the Italian experience and succeeded in achieving the quorum. The Greek government, through a circular by Minister of the Interior Michelakis, declared 'illegal' the self-organized water referendum threatening the organizers with arrest for "obstructing the electoral process" (Save Greek Water, 2014; EPSU, 2014).[2] The volunteers behind the referendum announced that they would go ahead risking arrest by the police. Nevertheless, the 11 local mayors of the metropolitan area supported the referendum and denounced in forceful terms the attitude of the government. A thirty-person international delegation, many of whom were affiliates of the European Federation of Public Services Unions (EPSU), came to Thessaloniki to observe the referendum (EPSU, 2014). None of the organizers were arrested. Instead, the referendum resulted in a 98 percent vote against water privatization. This citizen-led initiative mobilized 218,002 voters and sent a crystal-clear message rejecting the planned sale of EYATH shares to private investors. In this light, the referendum can be seen to be the biggest success of citizen-led mobilization against austerity and in support of public water. In spite of this success, the Greek government ignored the outcome.

Situating struggles against water privatization in Greece and Portugal within an understanding of the commons became crucial (Bieler and Jordan, 2016). Indeed, the citizens of the group K136 against water privatization in Thessaloniki viewed the crisis as an opportunity to intensify the search for democratic alternatives. Working on an alternative model of how to run the city's water services, it emphasized the importance of a new form of democratic water management based on direct democracy and self-management (Steinfort, 2014).

Theodoros Karyotis, one of the founders of K136, spoke of his experiences with the referendum saying:

> It was an empowering moment, where Thessalonikians felt that they have recovered a bit of the dignity taken away from them in four years of austerity and dispossession. Many people, disillusioned by the electoral process, went out just to vote in the referendum; it is unbelievable to see what great effect making one's voice heard on an important issue can have in a political system that systematically treats voters as clients and promotes apathy and resignation. This is another aspect where the referendum has been crucial: in creating political consciousness and collective empowerment.
>
> (Karyotis, 2014)

Karyotis warned against crying victory over the success of the referendum:

> Staying humble at this moment means: Recognizing that the movement is diverse and multitudinous; that no one person or group can represent or speak on behalf of the whole movement; that no one political party, mayoral candidate or group can claim credit for the outcome of the referendum; and most importantly, that the big common "NO" to the privatization is only a preamble to an open and democratic discussion about the future of water management and about the best possible way to ensure democratic participation, environmental protection, transparency and social justice in the provision of this valuable resource.
>
> (Karyotis, 2014)[3]

Who wins, who loses? (and what?)

The Greek government at first declared the referendum illegal although the organizers had the support of all eleven municipalities in the service area of EYATH in Thessaloniki. When the government rejected the outcome of the referendum, the EYATH water workers' union decided to go to court. In response, the Council of State decided that the government was not allowed to sell the water companies of Athens and Thessaloniki because such an act contravened the Greek Constitution: water and sewerage services were inseparable from the core State activity (Katrakaza, Papoutsi and Xenoyiannis, 2016). In the end, the government decided to put a stop to the privatization of water services in both Thessaloniki and Athens (MacroPolis, 2014).

A remarkable aspect of the referendum was that it had the support of the local governments and local politicians across the political spectrum. Such an organization was remarkable in spite of the opposition of national governments. Here we see that local authorities appear to develop a closer connection to people in understanding their needs and daily struggles during the time of the crisis, as opposed to a national State subject to elite capture and dominated by two political parties that had ruled Greece for over twenty years. There is also a difference in the interests between the local and national government. As Bieler shows in his chapter on "structure and agency in the struggle for public water in Europe", the national government is merely "obeying Troika orders" by imposing the sale of public water companies and slashing public services (Bieler and Jordan, 2017). Local government, however, has to provide these services, face local inhabitants and local public servants every day, witnessing the impact of austerity measures on local peoples while maintaining the trust and legitimacy of their political authority granted by local constituencies. Whereas Bieler sees the class struggle and the struggle of capital against labor, we focus in the case of Thessaloniki on the conviction of the inhabitants of the city that water should be a common and public good. Commoditization and privatization of water involved, both in the eyes of the movements in Thessaloniki as well as in the Right2Water movement, crossing a line that should not be crossed. This was confirmed by the outcomes of the Referendum and of the European Citizens' Initiative (van den Berge et al., 2018). Such proposals, therefore, led to a mass mobilization that crossed political lines and conventional political groups. There is however a clear aspect of class struggle in the sense that the referendum helped in creating collective empowerment and raising the voice of workers and people while linking the water movement to the wider social movement that was rising.

Soon after, the Greek government had to call for new elections as it could not achieve support from the Greek population for another deal with the Troika on new austerity measures. This moment turned into a landslide defeat for traditional governing parties followed by the rise of the 'extreme' right wing and the two left-wing parties New Democracy and Syriza. The latter became the winner of the elections, gaining the support of the population to negotiate another deal with the Troika. Although it was said that conditions imposed by the Troika were 'too high', others argued that the new government had to give in to reach a new bailout. This apparent capitulation was a huge disappointment to the movements that had put their hope in the Syriza government. Syriza, during the elections, had promised a clear social policy to the people suffering from austerity measures imposed and from decades of governmental mismanagement. Before the election, the party had campaigned with social movements against privatization, but once in government, it changed its position. The Syriza government agreed to a new deal with the Troika that included the sale of (23 percent of) the water company EYATh, plus management. Trying to downplay the sale by pledging that 51 percent would remain in government hands, the government were confronted by organizers of the referendum who contested any further privatization (Archontopoulos, pers. comm, 2018). Again, disappointment was the result, a situation that can be explained by the structuring conditions of capitalism (see the chapter by Bieler in this volume), conditions that the new Syriza government were not able to evade.

The Thessaloniki water company EYATH, however, did put its ideas on social policy into practice. It introduced social tariffs that allowed poor people to receive 30 m^3 water free of charge for a four-month period. This applied to low-income households with total taxable income less than 8.000 EUR per year, increased by 3.000 EUR for each dependent child. For consumption of 31 to 80 m^3 of water per four-month period, they were able to benefit from a 50 percent discount of their water bill.[4]

A year after the referendum on water privatization in Thessaloniki, the President of the Greek Parliament Zoi Konstantopoulou solemnly pledged her support for the implementation of the Human Right to Water as a just societal demand. She made the announcement at a meeting of the Thessaloniki City Council, explaining that she wanted Greece to become the first EU-country to recognize such a right. The President of the Parliament declared that she is open to the demands of social movements and supported the Right to Water movement.[5]

At the European level Right2Water achieved a wonderful result with the first ever successful ECI. However, the European Commission did not propose new legislation (see van den Berge et al., 2018). And in Greece, the referendum did not lead to the expected new legislation to prevent the privatization of the water companies. On a more positive note, the movements considered the increased awareness around water among citizens as perhaps the best outcome. In 2015, over 50 citizen groups – made up of thousands of people – therefore supported the struggle to make water a public good (Steinfort, 2014).

In 2015, for the first time in modern Greek history, the two traditional parties that had ruled the country since the end of the military dictatorship were cast out of government. This outcome did not bring the political change that people – especially the poor, lower class workers and the EYATH water workers – had hoped for. Instead, they paid the price of the crisis and remained losers in the fight against austerity claiming success in preventing the sale of the most essential public service providers. The European Commission, as part of the Troika, above all lost credibility. Insisting that it had not forced privatization upon Greece, the evidence was however clear. In a letter to Food and Water Europe (which is part of the Right2Water movement), the Commission admitted its support for privatization.[6] While temporarily losing in its push for full privatization of Greek state assets, the EC was successful in ensuring the Greek government's acquiescence towards the Troika in its demands for a third bailout.

Forces of capital continue to push back against the gains made by social movements. The third bailout agreement between Greece and the Troika in July 2015 included provisions for further privatization of the Thessaloniki and Athens water companies (Rettman, 2015; Pempetzoglou & Patergiannaki, 2017). The bailout agreement, therefore, outlined the need to establish "a new independent fund (the 'Super Fund'), which will have 'in its possession valuable Greek assets" (European Commission, 2015: 28). The two biggest water companies in Greece were included in this new fund. In the visit of President Macron to Greece, a delegation from the French multinational Suez took part and expressed their continuing interest in EYATH. While the Super Fund was supposed to raise the Greek income in 2018, the Greek water movements were preparing for a new campaign against privatization and a continuing struggle. In this campaign, some activists found themselves pitted against some of their previous supporters who, after the referendum, had joined Syriza. K136 is no longer a member of the new campaign. Instead, they stuck to their ideal of a cooperative and to the principle of self-management, something that does not match with the ideal of a public and state-owned company. The movements are now divided along different ideological lines between 'cooperativists' and 'public ownership activists.'

Discussion and conclusions

The ECI, based on a broad alliance of trade unions, social movements and NGOs, was successful at a time, when austerity policies were enforced across the EU member states, including pressures towards further privatization especially on the countries in the EU's periphery such as Greece. Although the Greek people were facing many difficulties in their daily lives (huge and increasing unemployment, closure of banks, non-payment of salaries, etc.), they supported the Right2Water from the moment that they saw the link between EU policy and (national) privatization plans.

The simultaneous development of the ECI Right2Water and the imposition of austerity measure by the Troika pushed social movements at various levels into each other's arms. The European campaigners found support for their cause at local levels in Greece, and Greek activists realized that their protest should not only be directed to their national government but also needed to resonate in the European capital. A situation of mutual alignment of forces and arguments emerged with Greek popular support for the Right2Water in exchange for European popular support to SOSteToNero. Working on the same campaign simultaneously enabled links to be established across borders, and in turn, facilitated international support for local campaigns. The anger over severe austerity, as well as the conviction that 'water is life', made people stand up against the privatization of water. The confidence that SOSteToNero had gained over the previous years was paid back in huge support. After a large turn-out and significant rejection of privatizing the Thessaloniki water company in this 'unofficial' referendum, with 98 per cent voting 'NO', the pressure on the Greek government not to privatize mounted. In a parallel but entwined process, a sufficient number of signatories in Greece was achieved for the ECI, despite the high bureaucratic barriers to achieving a quorum.

Similarly, SOSteToNero and the other organizers of the referendum on water in Thessaloniki politicized mobilized citizens against the privatization of water service provision, showing that a different form of managing a water utility is possible. Both movements achieved and sustained their success by keeping a diverse range of people acting for the same purpose. The social movements managed to unite these diverse interests for a common goal (van den Berge et al., 2018), however, success is fragile if victory is called too early and a movement can fall apart when individuals seek their own gain. Water appeared a powerful agency in the struggle for social justice and in unifying different social groups. At the same time, it is no more than a component in the struggle, having its limitations for achieving wider change in capitalist society as shown in the chapter by Bieler (within this volume). Nevertheless, neither Right2Water nor SOSteToNero, K136 and other Greek movements had the intention of fighting the political order. They opposed the privatization of water and in this fight, they succeeded.

After the success of the referendum and Syriza winning the elections, the disappointment regarding Syriza's lack of commitment to preventing the privatization of the water companies made social movements and activists turn their back on the 'old' political fractions and caused the falling apart of water activists into historical-political divisions. For the new movement, it should be a lesson that unity among the diverse is a condition for success in social struggles. Nevertheless, importantly, both the ECI as well as the referendum have led to greater awareness and consciousness about the importance of water services in the daily life of Greece, and consequently the growing willingness to resist privatization. Recent developments in Greece show that this struggle against privatization must continue.

Notes

1 www.136.gr/article/what-initiative-136
2 www.savegreekwater.org/archives/4080 and www.right2water.eu/news/right-water-condemns-greek-government-obstructing-water-referendum
3 http://www.autonomias.net/2014/05/people-vs-corporate-rule-some-personal.html
4 www.eyath.gr/swift.jsp?CMCCode=060202&extLang=LG
5 http://rights4water.net/en/articles/38-international-conference-on-water-privatization
6 www.foodandwatereurope.org/pressreleases/eu-commission-forces-crisis-hit-countries-to-privatise-water/

References

Archontopoulos, Y. (2018) *Personal interview at Water Workers Union office – EYATH, Thessaloniki*. 15 February.
Bakker, K. (2010) *Privatizing water: Governance failure and the world's urban water crisis*. Ithaca, NY: Cornell University Press.
Bieler, A. (2015) Mobilising for change, the first successful European Citizens' Initiative 'Water is a Human Right!'
Bieler, A. and Jordan, J. (2016) 'Mobilising against austerity: Greek and Portuguese labour in the resistance against water privatisation'. 3rd International Sociological Association (ISA) Forum of Sociology, Vienna, Austria, July 10–14.

Bieler, A. and Jordan, J. (2017) "Commodification and 'the commons': The politics of privatising public water in Greece and Portugal during the Eurozone crisis", *European Journal of International Relations*, 24 (4), pp. 1–24.
Boelens, R., Perreault, T. and Vos, J. (2018) *Water justice*. Cambridge: Cambridge University Press.
Calvário, R., Velegrakis, G. and Kaika, M. (2016) 'The political ecology of austerity: An analysis of socio-environmental conflict under crisis in Greece', *Capitalism Nature Socialism*, 28 (3), pp. 69–87.
European Commission (2010, May) 'The economic adjustment programme for Greece', *European Economy*, Occasional Papers 61.
European Commission (2011) "Regulation (EU) No 211/2011 of the European parliament and of the council of 16 February 2011 on the citizens' initiative", *Official Journal of the European Union*, L 65, 11.3.2011.
European Commission (2015, August 19) "Memorandum of understanding between the European commission acting on behalf of the European stability mechanism and the Hellenic republic and the bank of Greece". Athens and Brussels. Available at: https://ec.europa.eu/info/sites/info/files/01_mou_20150811_en1.pdf
European Commission Directorate-General for Economic and Financial Affairs (2010, May) *The economic adjustment programme for Greece* (European Economy, Occasional Papers 61). doi:10.2765/46750
European Commission Directorate-General for Economic and Financial Affairs (2011, July) *The economic adjustment programme for Greece, fourth review – Spring 2011* (European Economy, Occasional Papers 82).
European Federation of Public Services Unions (EPSU) (2014) "Right to water condemns Greek government for obstructing water referendum!" Available at: www.epsu.org/article/right-water-condemns-greek-government-obstructing-water-referendum
EYATH (2016) 'Investor information – financial data – brief financial statements'. Available at: www.eyath.gr/swift.jsp?CMCCode=06060201&extLang=LG
Fray, K. and Romei, V. (2015) "A modern Greek tragedy: The crisis years in context", *Financial Times*, June 19. Available at: www.ft.com/content/bbd4a2c2-d43e-3d0c-a7a2-2917ff544dc2
Hall, D.J. and Lobina, E. (2012) *Water companies and trends in Europe 2012*. London: Public Services International Research Unit.
Kaika, M. (2017) 'Between compassion and racism: How the biopolitics of neoliberal welfare turns citizens into affective idiots', *European Planning Studies*, 25 (8), pp. 1275–1291.
Karyotis, T. (2014) 'People vs. corporate rule: Some personal notes from my participation in Thessaloniki's great #vote4water referendum', May 19. Available at: www.autonomias.net/2014/05/people-vs-corporate-rule-some-personal.html
Karyotis, T. (2017) 'The right to the city in an age of austerity', *Roar Magazine*, Issue #6: The city rises, Summer 2017. Available at: https://roarmag.org/issues/the-city-rises
Kishimoto, S. and Hoedeman, O. (2015) 'Leaked EU memorandum reveals renewed attempt at imposing water privatisation on Greece'. *Transnational Institute*, August 24. Available at: www.tni.org/en/article/leaked-eu-memorandum-reveals-renewed-attempt-at-imposing-water-privatisation-on-greece
Kretsos, L. (2011) 'Grassroots unionism in the context of economic crisis in Greece', *Labor History*, 52 (3), pp, 265–286.
Lapavitsas, C., Kaltenbrunner, A., Lindo, D., Michell, J., Painceira, J.P., Pires, E., Powell, J., Stenfors, A. and Teles, N. (2010) *Eurozone crisis: Beggar thyself and thy neighbour*. London: Research on Money and Finance.

Lawson, T. (2015) 'Reversing the tide: Cities and countries are rebelling against water privitazation, and winning', *Truth-Out*, September 25. Available at: www.truth-out.org/news/item/32963-reversing-the-tide-cities-and-countries-are-rebelling-against-water-privatization-and-winning

MacroPolis (2014) 'Greece shelves water privatisation plans, leaving gap in revenue targets', July 2. Available at: www.macropolis.gr/?i=portal.en.economy.1331

Menzel, K. (2014) "Council of state blocks Athens water privatization", *Greek Reporter*, May 26. Available at: https://greece.greekreporter.com/2014/05/26/council-of-state-blocks-athens-water-privatization

Mylonas, Y. (2014) 'Crisis, austerity and opposition in mainstream media discourses of Greece', *Journal of Critical Discourse Studies*, 11 (3), pp. 305–321.

Organisation for Economic Co-operation and Development (2017) 'Real GDP forecast (indicator)'. Available at: doi:10.1787/1f84150b-en

Pempetzoglou, M. and Patergiannaki, Z. (2017) 'Debt-driven water privatization: The case of Greece', *European Journal of Multidisciplinary Studies*, 5 (1), pp. 102–111.

Pentaraki, M. (2013) "If we do not cut social spending, we will end up like Greece: Challenging consent to austerity through social work action", *Critical Social Policy*, 33 (4), pp. 700–711.

Perreault, T., Boelens, R. and Vos, J. (2018) 'Conclusions: Struggles for justice in a changing water world', in Boelens, R., Perreault, T. and Vos, J. (eds) *Water justice*. Cambridge, MA: Cambridge University Press, pp. 346–360.

Rankin, J. and Smith, H. (2015) 'The great Greece fire sale', *The Guardian*, July 24. Available at: www.theguardian.com/business/2015/jul/24/greek-debt-crisis-great-greece-fire-sale

Rettman, A. (2015) 'Greece to sell water, energy firms under EU deal', *EUObserver*, August 20. Available at: https://euobserver.com/economic/129936

Save Greek Water (2014) 'Greek Ministry of Interior threatens of arrests over water referendum and government "slides" on democratic deficit'. Available at: www.savegreekwater.org/archives/4080

Steinfort, L. (2014) 'Thessaloniki, Greece: Struggling against water privatization in times of crisis', *Transnational Institute*, June 3. Available at: www.tni.org/en/article/thessaloniki-greece-struggling-against-water-privatisation-times-crisis

Swyngedouw, E. (2005) 'Dispossessing H2O: The contested terrain of water privatization', *Capitalism Nature Socialism*, 16 (1), pp. 81–99.

United Nations General Assembly (2010) *Resolution A/RES/64/292: The human right to water and sanitation*. Available at:www.undocs.org/A/RES/64/292

van den Berge, J., Boelens, R. and Vos, J. (2018) 'Uniting diversity to build Europe's water movement Right2Water', in Boelens, R., Perreault, T. and Vos, J. (eds) *Water justice*, Cambridge, MA: Cambridge University Press, pp. 226–245.

Wainwright, H. (2013) 'Resist and transform: The struggle for water in Greece'. Avaialable at: www.redpepper.org.uk/tapping-the-resistance-in-greece

Wainwright, H. (2014) *The tragedy of the private, the potential of the public*. Amsterdam: Transnational Institute.

Zacune, J. (2013) *Privatising Europe: Using the crisis to entrench neoliberalism* (Working paper). Amsterdam: Transnational Institute.

13 Race, austerity and water justice in the United States

Fighting for the human right to water in Detroit and Flint, Michigan

Cristy Clark

Introduction

In April 2014, residents of Flint, Michigan began to complain about the water coming from their taps, and tens of thousands of residents in nearby Detroit had their water supply disconnected. Over the following two years, both cities experienced what can only be described as a water crisis and residents responded by campaigning forcefully for the human right to water. This chapter seeks to provide some historical context to the Flint and Detroit water crises and community responses, to highlight similarities with water justice campaigns elsewhere in the Global South, and to explore the significance of urban financialization and countervailing claims to the right to the city.

Water poverty is an issue we associate with the Global South, which may partly explain why the water crises in both Flint and Detroit, Michigan made international headlines (UN News, 2014; Felton, 2016; March, 2016; Palmer, 2016). But both crises actually occurred in the Global South and, indeed, demonstrate the nuanced meaning of the term. In addition to referring to the older geographic division between what were once called the First and the Third Worlds, the terms Global North and Global South also refer to the disparities of wealth within countries (Bello, 2004, p. 56).

Income inequality has increased markedly over the last forty years, particularly in urban areas, and this has resulted in a growing global divide between rich and poor communities, including within the United States (U.S.) (UN Habitat, 2016, pp. 69–84). And this inequality is reflected in more than just income – with substantial disparities in access to education, opportunities, services, political influence, and justice. This toxic mix of systematic disadvantage helps to explain what happened in both Flint and Detroit, and why socio-economic rights, like the human right to water, are becoming an increasingly important focus of social movements in the U.S.

Background to the crises

On the night of 23 July 1967, a race rebellion erupted in Detroit. It raged for five nights, and by its conclusion an estimated 43 people were dead and 696 wounded (Thompson, 2017, p. ix). While often described as a riot, this was a rebellion against structural and systematic racism. Thompson argues:

> It was the result of the grinding poverty that continued to exist in Detroit's black neighborhoods as white Detroiters enjoyed unprecedented prosperity. It happened because access to everything from good housing stock to strong schools remained elusive for black Detroiters while, for white city residents, such access was a given.
>
> (Thompson, 2017, p. ix)

Although the political elite had started paying lip service to equality through the introduction of a 'war on poverty' and formal equality measures like the *Civil Rights Act* of 1965 (which enabled the election of the City's first black Mayor in 1973), the 1967 Detroit Rebellion was a demand for substantive equality (Thompson, 2017, pp. xi–xv). But this was a bridge too far for America's white elite. As Lyndon Johnson reportedly said in 1957:

> These negroes, they're getting pretty uppity these days and that's a problem for us since they've got something now they've never had before, the political pull to back up their uppityness. Now we've got to do something about this, we've got to give them a little something, just to quiet them down, not enough to make a difference.
>
> (Caro, 2009, p. 955)

The most substantial political response to this so-called uppityness was a 'war on crime' that focused on *quieting them down* through over policing, criminalization, and mass incarceration (Thompson, 2017, pp. xi-xv).

Meanwhile, the white population of Detroit was leaving en masse to Wayne County's suburbs. Between 1970 and 2010 the city's population shrunk from 1,514,063 to 713,777 (Thompson, 2017, p. xv). The combined effects of white flight and mass incarceration served to empty the City of people and any meaningful tax base. When coupled with de-industrialization and dramatic reductions in state revenue sharing, this ultimately led to economic collapse (Desan 2014). In March 2013, an Emergency Manager was appointed, and on 18 July 2013, faced with a debt of around $20 billion, the City of Detroit filed for Chapter 9 bankruptcy (*re City of Detroit*, 504 B.R. at 128). Early the following year, as part of new austerity measures, the City began disconnecting the water supply of tens of thousands of households who were behind on their bills (*Lyda v City of Detroit* (2014)).

In April 2014, just as households in Detroit were having their water disconnected, households in the nearby city of Flint were watching their water turn brown, and blue, green, beige and yellow (Sanburn, 2016, p. 34). In short, the 'water physically looked, tasted and smelled foul' (Michigan Civil Rights Commission, 2017, p. 12). Residents began experiencing painful skin rashes, respiratory problems and hair loss (Sanburn, 2016, p. 34; Clark, 2018, pp. 79–

80). In time, the evidence mounted that their tap water, which had recently been switch over to the Flint River to reduce costs, was causing acute sickness – lead poisoning and a deadly outbreak of Legionnaires' disease (Flint Water Advisory Task Force, 2016, pp. 6, 16).

The City of Flint, in Michigan's Genesee County, is less than an hour's drive from Detroit and shares a very similar history – including a past reliance on the automobile industry, stark racial inequalities, and economic collapse caused by de-industrialization and 40 years of white flight (Michigan Civil Rights Commission, 2017, pp. 57–58). Flint joined Detroit in the race rebellion of July 1967, with similarly anaemic results in terms of structural change (Michigan Civil Rights Commission, 2017, pp. 65–67). And, like Detroit, Flint's recent economic woes led to the imposition of Emergency Management and austerity measures, resulting in even more dramatic consequences for its water services.

Up until April 2014, both cities shared a water service – the Detroit Water and Sewerage Department (DWSD), which pumped and treated fresh water from Lake Huron and piped it to households and businesses throughout Wayne and Genesee Counties. In 2012, as Flint's lease with DWSD was coming up for renewal, the City's Emergency Manager was considering alternative options (Hammer, 2016, p. 14). What happened next is complex (and the subject of multiple lawsuits), but by 2013 Flint's Emergency Manager had committed the City to supporting the construction of a brand-new private pipeline that would run *parallel* to the DWSD pipeline and transport untreated Lake Huron water into Genesee County (Flint Water Advisory Task Force, 2016; Hammer, 2016). This raw water would meet the (largely white) suburban desires of independence and be cheaper for agriculture and businesses in Genesee County but would require Flint to upgrade its aging water treatment plant (WTP) at an estimated cost of $61 million (Hammer, 2016, p. 16).

As Flint didn't have any money available to upgrade its WTP or to service its debt for the new pipeline project, the City's managers came up with a workaround – they would disconnect from DWSD at a saving of some $11 million annually and rely on water from the (heavily polluted) Flint River for two years while the new pipeline was being built (Flint Water Advisory Task Force, 2016, p16s). From these savings, around $8 million was used to upgrade the Flint WTP – a shortfall of $53 million from the estimated amount required to make the water safe (Hammer, 2016, p. 33).

Flint's decision to disconnect from DWSD was also bad news for the people of Detroit as it resulted in an annual loss of some $22 million from the system and created pressure to increase rates for remaining customers (Devitt, 2013). The mass disconnections in April and May of 2014 were no coincidence.

Emergency management and the finacialization of urban governance

Discussions around the realization of the human right to water often focus on the financial details of good water governance – *efficiency, sustainability, technical capacity* (see, eg, Cardone & Fonseca, 2003; Winpenny, 2003) – and the normative content of the right – *access, affordability, sufficiency* (see, eg, Winkler, 2012). But, as is so often the case, what happened in Flint and Detroit had far more to do with more fundamental issues of good governance – *transparency, participation, accountability* – and general principles of human rights – *equality, non-discrimination, non-retrogression*. The technical details that underpin these governance failures and violations of the right to water merit attention, because they reflect a broader trend in urban governance both in the U.S. and globally.

In the late 1980s, Harvey (1989, p. 12) warned of the dangerous consequences of a shift towards 'urban entrepreneurialism' – a manifestation of the crisis of Keynesian Fordism. This entrepreneurial style of governance was characterized by inter-urban competition exerting an 'external coercive power over individual cities to bring them closer into line with the discipline and logic of capitalist development', and the promotion of public-private partnerships, which support speculative debt-driven investment underpinned by the absorption of risk by the public sector (Harvey, 1989, pp. 7, 10–11). The results of this shift included increased financial instability, indebtedness, the polarization of wealth, and the impoverishment and disempowerment of an expanded urban underclass (Harvey, 1989, p. 12). Peck and Whiteside (2015, p. 238) argue that over the intervening decades this entrepreneurialism has been compounded by the *constitutive financialization* of the city – 'symbolized by the atrophy of redistributive financing and the explosive growth of the municipal bond market.' This has led to predictable financial crises, structural adjustment by financial means, and 'a drift towards postdemocratic technocratic management' (Peck and Whiteside, 2015, p. 238).

The imposition of Emergency Management in both Detroit and Flint is perhaps the ultimate expression of this drift and it played a significant role in the development of both water crises. Peck and Whiteside (2015, p. 254) describe the emergency manager laws in Michigan – 'on most accounts the nation's strictest' – as a mechanism for installing 'preemptive, unitary, and close-to unilateral forms of financial control, overriding the powers of elected officials and circumventing local democratic channels.' The 'tsar-like' scope of the powers granted to Detroit and Flint's Emergency Managers was intensified by the fact they bore 'no electoral accountability, channelling interests that effectively trumped those of local citizens' (Peck and Whiteside, 2015, p. 254). As a result, the health and safety of residents effectively ceased to be a priority for both administrations, and residents were completely stripped of their capacity to participate in decision making or to hold their government to account for its failures. As the Michigan Civil Rights Commission (2017, p. 113) documented in their report, Flint residents felt they were 'no longer heard when an emergency manager is in place.'

Peck and Whiteside (2015, p. 255) liken emergency management measures to the 'structural-adjustment approaches pioneered by multilateral lending institutions like the World Bank, not least in their apparently wilful contempt for local democracy.' This is also reflected in a range of other similarities between

Detroit and Flint's water crises and those experienced by cities in the Global South under structural adjustment and other similar measures. In South Africa, for example, residents of Soweto in Johannesburg had their water disconnected via the technocratic means of prepayment meters, which were imposed on residents without consultation (see, eg, Clark, 2012). When these disconnections were challenged in court under South Africa's constitutionally protected human right to water, the City of Johannesburg focused heavily on issues of financial sustainability, arguing the move was necessary to balance its budget (City of Johannesburg Answering Affidavit in *Mazibuko v The City of Johannesburg* High Court of South Africa (Witwatersrand Local Division) (2007) at para 25.6–25.8). In a concerning demonstration of the hegemonic power of this financialized logic, the Constitutional Court of South Africa upheld the legality of the City's actions by focusing on the technical and financial complexity of the City's policy dilemmas rather than the rights of Soweto residents (Wilson and Dugard, 2014, p. 56).

Equality

The apparent urgency of each City's financial imperatives was underpinned by the same logic in Johannesburg, Flint and Detroit, but it was also exacerbated by another similar factor: the construction of a pipeline for the benefit for agribusiness and wealthy white suburbs. Johannesburg's water supplies are boosted by water from the controversial Lesotho Highlands Water Project, which was conceived during the apartheid era and constructed in post-apartheid South Africa with $8 billion in debt financing (Bond, 2002). While the impacts on affected communities in Lesotho have been severe (Mwangi, 2007), the ongoing costs of servicing the debt also exacerbated water affordability problems for poor black households in Soweto, thus contributing to the City's decision to impose prepayment meters (Bond, 2002, pp. 243–249). Just as Flint residents were denied access to clean water in order to cover the costs of a new water pipeline for the benefit of others, so too were the residents of Soweto. The common factor: race.

In its final report on the Flint water crisis, the Michigan Civil Rights Commission (2017, p. 93) found that it could be best understood as an instance of environmental racism – which 'occurs when people of color repeatedly suffer disproportionate risks and harms from policies and decisions that equally benefit all.' In her analysis of Flint, Ranganathan (2016, p. 18) cautions that it is important not to see this act of environmental racism as some kind of aberration. Instead, she argues that it is 'inextricable from the workings of liberalism, specifically racial liberalism as it took root in America's cities from the mid-20th century onwards' (Ranganathan, 2016, pp. 18–19). She further elaborates, 'our understanding of Flint's predicament—the slow poisoning of an entire generation of African-Americans—can be deepened if we read it as a paradigmatic case of racial liberalism's illiberal legacies' (Ranganathan, 2016, p. 19).

Indeed, Johannesburg, Detroit and Flint share a common history of explicitly discriminatory laws and policies that served to create and entrench racial disadvantage. Apartheid-era segregation is well known, but similar outcomes were achieved in the U.S. through a combination of state-sanctioned private power and the technical detail of the *Home Owners' Loan Act of 1933* (Michigan Civil Rights Commission, 2017, pp. 30–40). The origin of the term 'redlining', for example, can be traced back to these discriminatory housing policies when 'D rated' neighborhoods were coded red due to their ineligibility for federal housing loans (Michigan Civil Rights Commission, 2017, pp. 33–40). One of the key criteria for a D rating was the presence of non-white households.

These discriminatory housing (and related educational) policies had 'cumulative and compounding effects' by seriously limiting opportunities for investment, security and career choice (Michigan Civil Rights Commission, 2017, p. 42). The net result, as Ranganathan (2016, p. 25) summarises, 'was a consolidation of white wealth in the form of state-subsidized housing equity, which in turn translated into larger inheritances and better educational opportunities for the next generation.' As Jepson, Wutich, and Harris (this volume) point out, these educational inequities also had direct implications for the capacity of communities to obtain 'the knowledge and power needed to confront . . . water injustices.' For Flint residents, for example, this included repeatedly being denied access to information about lead contamination levels in their tap water.

Another legacy of this structural inequality is debt. Many households in Detroit and Flint have unpaid water bills due to their incapacity to pay the usually high tariffs imposed on city residents (Murthy, 2016, pp. 167–170; Michigan Civil Rights Commission, 2017, p. 54). In both cities these debts have resulted in household disconnection, but they have also been used by the City of Detroit to both coerce poor (mostly black) households into complying with draconian payment plans and to legitimize the City's policies by branding these payment plans as socially responsible affordability measures (Murthy, 2016, pp. 220–224). Municipal governments in South Africa have made similar use of household debts. In Soweto, for example, the high level of water debt was used as a key justification for the forced installation of prepaid water meters (*Mazibuko v City of Johannesburg* (2009) 28 ZACC), while Yates and Harris (2018) have recently documented a similar approach in Cape Town.

The long-term implications of this shared history of structural discrimination continue to reverberate today. There are similar patterns of spatial segregation in each city, which, along with the related endemic poverty and the disciplinary measures used in response to debt, serve to strip the black population of both financial and political power. A tangible outcome of this in all three cases was the violation of the human right to water.

Non-discrimination and non-retrogression

The U.S. government has customarily been hostile to the notion of enforceable socioeconomic rights and has not ratified the *International Covenant on*

Economic, Social and Cultural Rights 1966 ('ICESCR'). Despite this, water justice campaigners have been fighting for the domestic recognition of the human right to water and former U.S. critics of socioeconomic rights have more recently begun to embrace both the importance and enforceability of these rights (see, eg, Sunstein, 2004 citing Roosevelt's historical call for 'freedom from want'). As such, it worth considering how socioeconomic rights have been interpreted at the international level by the Committee on Economic Social and Cultural Rights ('CESCR') and how CESCR has responded to traditional critiques – including the assertion that socioeconomic rights are too vague to be individually enforced and that adjudication in the courts is institutionally inappropriate, because enforcement requires positive government action (usually implying expenditure and complex policy choices) (Sunstein, 1993;[1] Dennis & Stewart, 2004; Whelan, 2007).

CESCR has sought through its General Comments to respond to these criticisms by elaborating on the obligations imposed under the ICESCR in order to give more certainty and weight to the socioeconomic rights it protects. This elaboration has included the delineation of the obligations into three categories of duties: *respect, protect* and *fulfil* (CESCR, 1999, para. 15; CESCR, 2002, para. 20), and the identification of *obligations of immediate effect* – such as non-discrimination (contained in Article 2(2) of ICESCR) and the presumption against retrogressive measures (CESCR, 1990, paras. 1, 9). As the United Nations Office of the High Commissioner for Human Rights notes, although the obligation to *fulfil* may involve a claim on the public purse, the obligations to *respect* and *protect* will often involve so-called negative duties that do not have significant resource implications for the government (OHCHR, 2007, para. 9). This is also the case for the obligation of non-discrimination and the presumption against retrogressive measures, which are particularly relevant to the right to water and to the violations experienced in Johannesburg, Detroit, and Flint (described previously).

CESCR (2002, para. 14) emphasizes that the enforcement of the right to water includes the rigorous application of principles of non-discrimination, especially at the level of resource allocation and choice of service models. It also requires that special attention be paid to the rights of minorities and vulnerable populations when scrutinising related government decisions. This has implications for the pricing of water services and for the kind of water governance model that is compatible with the full realization of the right to water. A market-dominated model of water governance that focuses excessively on efficiency and financial sustainability risks, at best, ignoring issues of equity and non-discrimination, and, at worst, further entrenching inequitable patterns of water access that stem from the historic exclusion of the poor (see, eg, Birdsall & Nellis, 2003; Dagdeviren, 2008; Dugard, 2010). This is precisely what took place in Johannesburg, Detroit, and Flint.

The 'presumption against retrogressive measures' (CESCR, 1990, para. 9) is related to the obligation that States parties to the ICESCR continue to 'take steps' to realize the rights protected under the ICESCR (Article 2(1)) since this implies that governments must refrain from taking steps that would negatively affect this realization. In relation to the right to water, CESCR (2002, para. 19) has stated:

> If any deliberately retrogressive measures are taken, the State party has the burden of proving that they have been introduced after the most careful consideration of all alternatives and that they are duly justified by reference to the totality of the rights provided for in the Covenant in the context of the full use of the State party's maximum available resources.

Again, the water disconnections in Johannesburg and Detroit, and switching the water supply of Flint from a safe to an unsafe service, arguably breached this obligation of non-retrogression as they involved positive government action that negatively affected access to the right to safe and affordable water.

The appropriate role of the state

This delineation of three categories of duties owed by States parties to the ICESCR highlights the complexity of the role of the government in the realization of the human right to water. The active role of the City governments in breaching the right in all three cases described previously supports Dwinell and Olivera's (2014) caution against overly centering the state (be it local, provincial, or national) in any campaign for water justice. However, Angel and Loftus (2019, p. 3) are also right to emphasise that 'the state is not a "thing" but is rather an idea, an effect, or a reified set of social relations.' This relational understanding of the state (and the right to water) means that water justice activists must find a way to work 'at once in-against-and-beyond' both the state and the human right to water (Angel and Loftus, 2019, p. 5). In this volume, Jepson, Wutich, and Harris respond to this challenge by considering the human right to water through the lens of the capabilities approach. They argue that this 'perspective opens new visions of what a right to water is' by understanding 'individuals and communities as citizens and political actors rather than flattened as only consumers of water' (Jepson, Wutich, and Harris, this volume).

In both Flint and Detroit, water justice activists have worked 'at once in-against-and-beyond' both the state and the human right to water in their ongoing campaigns against cost-cutting, mass disconnections and excessive cost recovery policies (see e.g., Clark, 2017, pp. 251–252). In Lyda et al v City of Detroit and Others ('*Lyda*'), for example, Detroit residents challenged their water disconnections through an adversary complaint in the Detroit bankruptcy proceedings, claiming the City had breached their rights to both due process and equal protection. Although the complaint did not survive a motion to dismiss, it drew attention to the plight of residents and served to frame their demands against the state in human rights language.

Following the judgment in *Lyda*, civil society organizations invited the UN Special Rapporteurs on the human right to water and the right to housing to visit Detroit (OHCHR, 2014). In addition to documenting a lack of affordability and procedural justice, the Special Rapporteurs highlighted the high level of poverty and racial segregation in Detroit and the discriminatory effect of the mass disconnections (OHCHR, 2014). In 2016, UN Special Rapporteurs also issued a similar statement condemning the human rights violations in Flint:

> The Flint case dramatically illustrates the suffering and difficulties that flow from failing to recognize that water is a human right, from failing to ensure that essential services are provided in a non-discriminatory manner, and from treating those who live in poverty in ways that exacerbate their plight.
> (OHCHR, 2016)

This strategic decision by U.S. social movements to employ the language of the human right to water is noteworthy. Not only has the U.S. state been traditionally hostile to socioeconomic rights (as noted previously), but it abstained from voting on the 2010 United Nations General Assembly resolution on the right to water (see United Nations, 2010) and lacks explicit constitutional protection for the right. Viewed within this context, social movement engagement with the discourse of the human right to water (in *Lyda*, in the decision to engage with the UN Special Rapporteurs, and in broader campaign communications) can be seen as a radical strategy – one that challenges rather than reinforces the (liberal) status quo and one that frames the communities in Flint and Detroit as 'citizens and political actors' rather than 'only consumers of water' (Jepson, Wutich, and Harris, this volume).

Financialization, water and the right to the city

In the first edition of this book, Sultana and Loftus (2012, p. 8) asked how the right to water 'might acquire a material force within the world and how it might become actually world-changing.' The ultimately radical agenda of the water justice movements in Flint and Detroit, like the earlier radical agenda of water justice movements in Johannesburg, is a partial answer to this question. This is because a notable feature of the water justice campaigns in both the U.S. and South Africa was the sustained focus both on challenging the financialization of governance and the resultant democratic deficit, and on asserting a human right to water. While demonstrating the risk that legal engagement can serve to reinforce a hegemonic discourse by upholding the status quo (see eg, Bakan, 1997, p. 31), both campaigns also demonstrate that social movements are able to adopt a sophisticated approach to human rights litigation – framing acceptable legal claims in court while simultaneously pursuing more radical agendas (Clark, 2017).

The water justice campaign in Detroit, for example, is multifaceted and the decision to focus on the human right to water has taken place against the background of a broader, ongoing campaign to claim a more expansive right to the city (We the People of Detroit, nd). And they are having some tangible wins. In 2015, for example, Democratic State Government Representatives in Michigan proposed new legislation to better regulate water quality and affordability and to limit disconnections (Murthy, 2016, p. 230). Significantly, these American politicians justified their legislation on the basis of the human right to water. Even more significantly, the water justice movements in Detroit and Flint have been a catalyst for increased political activism across the board.

The struggles in Flint and Detroit to reclaim control over their cities are part of an ongoing tension between the forces of capital and community resistance that has played out throughout history. In the urban context, the most dramatic of these struggles resulted in the (short-lived) 1871 Paris Commune – a 'dramatic seizure of the government by Parisian workers [as] a response to smouldering class antagonisms' (Vasudevan, 2015, p. 321). In reflecting on the relevance of the Commune to the (ultimately unsuccessful) 1968 uprising in Paris, Lefebvre coined the concept of 'the right to the city' (Butler, 2012, pp. 34–35). He described the act of claiming the right to the city as 'autogestion' – a process of seizing control from below in order to collectively manage decisions and common resources (Purcell, 2013, p. 147). As Bond (2012, p. 198) has pointed out, '[e]ach different struggle for the right to the city is located within a specific political-economic context in which urbanization has been shaped by access to water.'

While it is easy to dismiss human rights claims as individualistic, the right to the city is better defined as a 'right to resistance ... built upon common demands and social cooperation' (Negri, 2004, p. 110). In the context of the U.S., these radical demands require the development of new visions of alternative social relations and a genuine belief that 'another world is possible.' The last decade has demonstrated a new commitment to this alternative vision with social movements like Occupy emerging seemingly out of nowhere.

As Lefebvre (1995, p. 348) contended, 'social change is driven by the play of a dialectic between the possible and the impossible.' Reflecting on this dialectical process of social change, Harvey (2013, p. xv) argues, 'the right to the city is an empty signifier. Everything depends on who gets to fill it with meaning.' As a result, '[t]he definition of the right is itself an object of struggle, and that struggle has to proceed concomitantly with the struggle to materialise it' (Harvey 2013, p. xv). The same could be said of the human right to water and, as Sultana and Loftus (2012, p. 9) have argued, 'the key challenge is to be able to fill this empty signifier with real political content.' Specifically, they ask, whether the right to water can mean 'the right to be able to participate more democratically in the making of . . . the "hydrosocial cycle"' (Sultana and Loftus 2012, p. 8, citing Linton, 2010; 2012; Swyngedouw, 2004).

After examining the empirical realities of grassroots struggles for the right to water – whether in Flint, Detroit, or Johannesburg – it seems the answer to this question is yes, particularly when linked to a larger struggle for the right to the city. As Bond (2012, p. 198) argues, this class-based right involves three components: 'equal participation in decision-making, equal access to and use of the city and equal access to basic services.' This emphasis on participation addresses the democratic deficit created by financialization and historical structural barriers to civic participation. Similarly, the focus on equality implicitly introduces a specific challenge to the liberal status quo, since, as Ranganathan (2016, pp. 20–21) emphasises, 'racial hierarchy is foundational—and not simply incidental—to the workings of capitalism and an ostensibly democratic, liberal market society.'

In places like Detroit and Flint, the tangible reality of water rights violations has proven to be a powerful motivator to come together, to resist the current order of things and to fight for inclusive governance that achieves substantive equality – in terms of access to both water and the city at large. The water crises in both cities were the result of systematic racism, entrenched poverty and a complete democratic deficit caused by the financialization of the cities' governments. These underlying causes – and the inevitable results – mirror a story that has played out over and over again in cities across the Global South and demonstrate that breaches of the human right to water have more to do with disparities of power than anything else.

Indeed, water justice activists have already heeded this lesson and have embedded their struggles for the right to water within a broader campaign to reclaim power from below. When embedded within broader (radical) claims for the right to the city, the right to water could well carry the seeds of genuine structural change. The visceral attachment people have to water – as a source of life, identity and community – makes it a dangerous frontier for capitalism's ongoing 'accumulation by dispossession' (Harvey, 2004), which in turn makes the ongoing struggles in Michigan worthy of further attention.

Note

1 Note that Sunstein appears to have changed his mind in more recent years (Sunstein, 2004).

References

Angel, J. and Loftus, A. (2019) 'With-against-and-beyond the human right to water', *Geoforum*, 98, pp. 206–213.
Bakan, J. (1997) *Just words: Constitutional rights and social wrongs*. Toronto, ON: University of Toronto Press.
Bello, W. (2004) 'The Global South', in Mertes, T. (ed) *A movement of movements: Is another world really possible?* London: Verso.
Birdsall, N. and Nellis, J. (2003) 'Winners and losers: Assessing the distributional impact of privatization', *World Development*, 31 (10), pp. 1617–1633.
Bond, P. (2002) 'A political economy of dam building and household water supply in Lesotho and South Africa', in McDonald, D., *Environmental Justice in South Africa*. Cape Town: Ohio University Press, pp. 223–269.
Bond, P. (2012) 'The right to the city and the eco-social commoning of water: Discursive and political lessons from South Africa', in Sultana, F. and Loftus, A. (eds), *The right to water: Politics, governance, and social struggles*. New York: Earthscan, pp. 190–205.

Butler, C. (2012) *Henri Lefebvre, spatial politics, everyday life and the right to the city*. Abingdon: Routledge.

Cardone, R. and Fonseca, C. (2003) *Financing and cost recovery*. Delft, Netherlands: International Water and Sanitation Centre.

Caro, R. (2009) *Master of the senate: The years of Lyndon Johnson* (2nd ed.). New York: Vintage Books.

Clark, A. (2018) *The poisoned city: Flint's water and the American urban tragedy*. New York: Metropolitan Books.

Clark, C. (2012) 'The centrality of community participation to the realisation of the right to water: the illustrative case of South Africa', in Sultana, F. and Loftus, A. (eds) *The right to water: Politics, governance and social struggles*. New York: Earthscan, pp. 174–189.

Clark, C. (2017) 'Of what use is a deradicalized human right to water?', *Human Rights Law Review*, 17 (2), pp. 231–260.

City of Johannesburg Answering Affidavit in *Mazibuko v The City of Johannesburg* High Court of South Africa (Witwatersrand Local Division) (2007).

Committee on Economic Social and Cultural Rights (1990) *General Comment No. 3: The nature of Sates parties obligations (Art. 2, Para. 1, of the Covenant)*, 5th CESCR session, UN Doc E/1991/23.

Committee on Economic Social and Cultural Rights (1999) *General Comment No. 12 on the right to adequate food (Art.11)* 20th CESCR session, UN Doc E/C.12/1999/5.

Committee on Economic Social and Cultural Rights (2002) *General Comment No. 15 on the Right to Water*, 29th CESCR session, Agenda item 3, UN Doc E/C/12/2002/11.

Dagdeviren, H. (2008) 'Waiting for miracles: The commercialization of urban water services in Zambia', *Development and Change*, 39 (1), pp. 101–121.

Dennis, M.J. and Stewart, D.P. (2004) 'Justiciability of economic, social, and cultural rights: should there be an international complaints mechanism to adjudicate the rights to food, water, housing, and health?', *American Journal of International Law*, 98 (3), pp. 462–515.

Desan, M.H. (2014) 'Bankrupted Detroit', *Thesis Eleven*, 12 (1), pp. 122–130.

Devitt, C. (2013) 'Flint departure a blow to Detroit water system', *The Bond Buyer*, April 18.

Dugard, J. (2010) 'Can human rights transcend the commercialization of water in South Africa? Soweto's legal fight for an equitable water policy', *Review of Radical Political Economics*, 42 (2), pp. 175–194.

Dwinell, A. and Olivera, M. (2014) 'The water is ours damn it! Water commoning in Bolivia', *Community Development Journal*, 49 (S1), pp. i44–i52.

Felton, R. (2016) 'Flint's water crisis: What went wrong?', *The Guardian*, January 17. Available at: www.theguardian.com/environment/2016/jan/16/flints-water-crisis-what-went-wrong

Flint Water Advisory Task Force (2016) *Final report*. Available at: http://flintwaterstudy.org/2016/03/flint-water-advisory-task-force-final-report

Hammer, P. (2016) *The Flint water crisis, KWA and strategic-structural racism* (Wayne State University Law School Research Paper No. 2016-17). Available at: https://papers.ssrn.com/sol3/papers.cfm?abstract_id=2812171

Harvey, D. (1989) 'From managerialism to Entrepreneuralism: The transformation in urban governance in late capitalism', *Geografiska Annaler B*, 71 (1), pp. 3–17.

Harvey, D. (2004) 'The "new" imperialism: Accumulation by dispossession', *Socialist Register*, 40, pp. 63–87.

Harvey, D. (2013) *Rebel cities*. London and New York: Verso.
Lefebvre, H. (1995) *Introduction to modernity: Twelve preludes September 1959–May 1961* (J. Moore, trans.). London: Verso.
Linton, J. (2010) *What is water? The history of a modern abstraction*. Vancouver, BC: University of British Columbia Press.
Linton, J. (2012) 'The human right to what? Water, rights, humans and the relation of things', in Sultana, F. and Loftus, A. (eds) *The right to water: Politics, governance and social struggles*. New York: Earthscan, pp. 45–60.
Lyda et al. v City of Detroit and Others No. 13–53846 (Bankr. E.D. Mich. September 29, 2014).
March, S. (2016) 'The human toll of Flint's water crisis', *ABC News*, March 12. Available at: www.abc.net.au/news/2016-03-13/flint-michigan-water-crisis-human-toll/7221924
Mazibuko v City of Johannesburg (2009) 28 ZACC.
Michigan Civil Rights Commission (2017) *The Flint water crisis: Systematic racism through the lens of Flint*. Available at: www.michigan.gov/documents/mdcr/VFlintCrisisRep-F-Edited3-13-17_554317_7.pdf
Murthy, S. (2016) 'A new constitutive commitment to water', *Boston College Journal of Law & Social Justice*, 36 (2), pp. 159–233.
Mwangi, O. (2007) 'Hydropolitics, ecocide and human security in Lesotho: A case study of the Lesotho Highlands Water Project', *Journal of Southern African Studies*, 33, (1), pp. 3–17.
Negri, A. (2004) *The porcelain workshop: For a new grammar of politics*. Los Angeles, CA: Semiotext(e).
Office of the High Commissioner for Human Rights (2007) *Report of the High Commissioner to the 2007 substantive session of ECOSOC (dedicated to the issue of progressive realization of economic, social and cultural rights. ECOSOC substantive session of 2007*, UN Doc E/2007/82.
Office of the High Commissioner for Human Rights (2014) 'Joint Press Statement by Special Rapporteur on adequate housing as a component of the right to an adequate standard of living and the right to non-discrimination in this context, and Special Rapporteur on the human right to safe drinking water and sanitation visit to city of Detroit', October 20. Available at: www.ohchr.org/EN/NewsEvents/Pages/DisplayNews.aspx?NewsID=15188
Office of the High Commissioner for Human Rights (2016) 'Flint: "Not just about water, but human rights" – UN experts remind ahead of President Obama's visit', May 3. Available at: www.ohchr.org/EN/NewsEvents/Pages/DisplayNews.aspx?NewsID=19917&LangID=E
Palmer, G. (2016) 'Flint water crisis: Living one bottle of water at a time', *BBC News*, January 22. Available at: www.bbc.com/news/magazine-35376517
Peck, J. and Whiteside, H. (2015) 'Financializing Detroit', *Economic Geography*, 92 (3), pp. 235–268.
Purcell, M. (2013) 'Possible worlds: Henri Lefebvre and the right to the city', *Journal of Urban Affairs*, 36 (1), pp. 141–154.
Ranganathan, M. (2016) 'Thinking with Flint: Racial liberalism and the roots of an American water tragedy', *Capitalism Nature Socialism*, 27 (3), pp. 17–33.
re City of Detroit, 504 B.R. 191 (Bankr. E.D. Mich. 2013).
Sanburn, J. (2016) 'The toxic tap', *Time Magazine*, February 1, pp. 32–38.

Sultana, F. and Loftus, A. (2012) 'The right to water: Prospects and possibilities', in Sultana, F. and Loftus, A. (eds) *The right to water: Politics, governance and social struggles*. New York: Earthscan, pp. 1–18.

Sunstein, C.R. (1993) 'Against positive rights: Why social and economic rights don't belong in the new constitutions of post-communist Europe', *Eastern European Constitutional Review*, 2 (Winter), pp. 35–38.

Sunstein, C.R. (2004) *The second bill of rights: FDR's unfinished revolution and why we need it more than ever*. New York: Basic Books.

Swyngedouw, E. (2004) *Social power and urbanization of water: Flows of power*. Oxford: Oxford University Press.

Thompson, H.A. (2017) *Whose Detroit? Politics, labor and race in a modern American city* (2nd ed.). Ithaca, NY: Cornell University Press.

United Nations Human Settlements Programme (UN Habitat) (2016) *World Cities Report*. Nairobi, Kenya: United Nations Human Settlements Programme (UN Habitat).

United Nations (2010) 'General Assembly adopts resolution recognizing access to clean water, sanitation as human right, by recorded vote of 122 in favour, none against, 41 abstentions', July 28. Available at: www.un.org/press/en/2010/ga10967.doc.htm

UN News (2014) 'In Detroit, city-backed water shut-offs "contrary to human rights", say UN experts', October 20. Available at: https://news.un.org/en/story/2014/10/481542-detroit-city-backed-water-shut-offs-contrary-human-rights-say-un-experts

Vasudevan, A. (2015) 'The autonomous city: Towards a critical geography of occupation', *Progress in Human Geography*, 39 (3), pp. 316–337.

We the People of Detroit (nd) 'Our story'. Available at: https://wethepeopleofdetroit.com/about

Whelan, D.J. (2007) 'Unpacking a "violations approach" to protecting economic, social and cultural rights'. Annual Meeting of the American Political Science Associate, August 30–September 2, Chicago, IL.

Wilson, S. and Dugard, J. (2014) 'Constitutional jurisprudence: The first and second waves', in Langford, M., Cousins, B., Dugard, J. and Madlingozi, T. (eds) *Symbols or substance: The role and impact of socio-economic rights strategies in South Africa*. Cambridge: Cambridge University Press, pp. 35–62.

Winkler, I. (2012) *The human right to water: Significance, legal status and implications for water allocation*. Oxford and Portland, OR: Hart Publishing.

Winpenny, J., (2013) *Financing water for all: Report of the World Panel on Financing Water Infrastructure*. Marseille, France: World Water Council.

Yates, J. and Harris, L. (2018) 'Hybrid regulatory landscapes: The human right to water, variegated neoliberal water governance and policy transfer in Cape Town, South Africa, and Accra, Ghana', *World Development Journal*, 110, pp. 75–87.

14 Class, race, space and the "right to sanitation"

The limits of neoliberal toilet technologies in Durban, South Africa

Patrick Bond

Introduction: the "right to sanitation" without water

Two positions emerged in South Africa during the 2000s regarding sanitation policy and law, following the commitment to a "right of access to water" in the country's 1996 constitution (several years ahead of Uruguay, Honduras, Algeria, Bangladesh, Kenya, Sri Lanka and others). First, water rights should not only translate into a free basic water supply of 6000 litres per household per month but should implicitly include a flush toilet. In this reading, the word "sanitation" is defined by the Water Services Act of 1997 and its 2002 implementing regulation as a component of "water services," which include "potable water supply services and sanitation (sewage and wastewater) services" (Republic of South Africa, 2002, 11). Merriam-Webster (n.d.) defines sewage as "refuse liquids or waste matter usually carried off by sewers." Whether termed sewage or sewerage, water-borne sanitation is simply the logical means of "carrying off" faecal waste matter.

The second position is that a *constitutional* right to sanitation does not exist per se, and although a 2001 White Paper declares "Basic sanitation is a human right," this is merely rhetorical, with minimalist clarification: "Government has an obligation to create an enabling environment through which all South Africans can gain access to basic sanitation services" (Republic of South Africa, 2001, 5). While the 1997 Water Services Act defines sanitation as "the conditions or procedures related to the collection and removal of sewage and refuse," implying that sewage lines would take excrement off-site ("removal"), the 2001 White Paper – aimed at "mainly rural communities and informal settlements" – added a new term ("disposal") to the definition: "the principles and practices relating to the collection, removal or disposal of human excreta, household waste water and refuse as they impact upon people and the environment" (Republic of South Africa 2001, 5). At that point, the "minimum acceptable basic level of sanitation" had come to entail "a system for disposing of human excreta, household waste water and refuse, which is acceptable and affordable to the users, safe, hygienic and easily accessible and which does not have an unacceptable impact on the environment; and a toilet facility for each household."

Certainly, in this latter framing, there is no right on the part of residents to have water-borne sewage installed. So, if in terms of public policy, sewage is "removed" without water, into pits (albeit "improved") via dry toilets, then there is no violation of rights. A limited basic supply of potable water to a household may be facilitated by a yard tap and a "soakaway" in which wastewater is simply allowed to saturate the nearby soil. But in this reading, no sewage pipe and treatment system are necessary.

Endorsing the latter position, the South African Human Rights Commission's (2018, 3) interpretation of sanitation entails "A toilet or ventilated pit latrine, which is safe, reliable, environmentally sound, easy to keep clean, provides privacy and protection against the weather, well ventilated, keeps smells to a minimum and prevents the entry of flies and other disease-carrying pests." Moreover, ecological limits now help to define the sanitation service levels for residents to be supplied by impoverished municipalities – and even rich ones like Cape Town, which, in 2017–2018, suffered drought-related physical supply shortages limiting consumption to a maximum of 50 litres per capita per day (lcd). A typical South African toilet flush is 8–12 litres.

Disputes over sanitation do occur in the courts, but only in 2009 were water supply standards tested in the Constitutional Court. There, Sowetans lost the famous case *Mazibuko v Johannesburg Water* when they demanded 50 free lcd as a basic supply (not the 25 on offer) and an end to pre-payment meters in black neighborhoods (Bond, 2013). As a result, although the pros and cons of policy were somewhat more rigorously debated, the highest level of the judicial system proved ultimately useless as a site to demand or enforce more generous, pro-poor public policy.

How does this dispute over sanitation rights play out in South African municipalities? This chapter considers the ideal case, Durban, South Africa, whose post-apartheid policy and management have been the Third World's most celebrated. Durban reveals the concrete manifestations of the right to sanitation *not as a sewage and waste-water system,* but as on-site disposal. The critique, thus, takes us through divergent narratives: from mainstream water-sector celebration of the city's experimentation on low-income black bodies to their micro-resistances.

Durban's "perfect toilet"

In Durban, oft-praised municipal water managers have used the second approach to sanitation rights in order to legitimate a new form of discrimination, based on a combination of neoliberal management of public services, geographical excuses, tokenistic forms of social policy, and exceptional self-promotion. Durban has witnessed occasional periods of stress in which restrictions against water lawns and washing cars were imposed, but unlike Cape Town, there have never been such extreme regional physical shortages as to restrict water for personal hygienic needs.

The water and sewage pipes that criss-cross the city are fed mostly from the Inanda Dam, whose late-1980s construction displaced thousands of black residents in the dying years of apartheid (their partial land compensation was only won after lengthy struggles in 2015). Yet in spite of the dam's 243 million litre capacity, tens of thousands of low-income households in the close vicinity – the Valley of a Thousand Hills – are officially provided with no water to flush away excrement. (In the higher-income, formerly all-white "Upper Highway" suburbs above the Valley floor – including Hillcrest, Assagay, Kloof, Botha's Hill, Forest Hills, Gillitts, Waterfall and Winston Park – there is no such restriction on flushing, either through sewage lines or septic tanks.)

For most low-income Durban residents, only 200 litres per household per day were initially provided for consumption during the early 2000s through on-site "Jojo" tanks with very little pressure (simply gravity), through a weakly-pumped piping system. After extensive protests, this Free Basic Water allocation was extended to 300 l/h/d in 2009. Sanitation is provided to most low-income residents through non-flush modes, including even a continuation of the hated "bucket system" of apartheid-era services, in which "night soil" in out-house buckets was removed by municipal workers each morning, replaced by a clean bucket.

This policy reflects an artificial water scarcity, applied to low-income areas below which run myriad pipes drawing upon the vast municipal dam. It was endorsed by a *Guardian* journalist: "Ground zero for the quest to find the perfect toilet for the 21st century's needs may as well be Durban, South Africa" (Kaye, 2012). Lead philanthro-capitalist Bill Gates (2010) visited Durban and blogged enthusiastically not only about Ventilated Improved Pit-latrines (VIPs) but the Urine Diversion (UD) toilets which allow urine to collect in one "soakaway" removable box and solids in another. In the latter, sawdust or sand is meant to be added, so the faeces dry and pathogens die. On behalf of the municipality, Teddy Gounden, Bill Pfaff, Neil Macleod and Chris Buckley (2013) explained, "The householder is required to remove the contents, dig a hole and bury the contents on site."

As an NGO expert – Water Dialogues leader Mary Galvin (2017, 113) – recorded, VIPs were "becoming unusable "full ups," being extremely expensive or virtually impossible to empty because of their location, and facing user opposition." Instead of the municipality vacuuming out the excrement (as was meant to be the case for the VIPs), households themselves would empty the faeces chamber in the UD, while if the plot was large enough, the urine chamber would empty directly into a soakaway system. (At one point, even the urine was potentially collectable, for purchase by the municipality and in turn, drying into phosphate for fertiliser.) Beginning in 2003, more than 85,000 UDs were placed in Durban's semi-peripheral areas.

The economics of the "neoliberal loo," seen from above

As a "neoliberal loo" (Amisi, Bond, Khumalo and Nojiyeza, 2008), the UD was attractive to the city because it would not require new piping nor pumping water into the lower-income neighborhoods. Contrasting expectations for non-neoliberal sanitation policy were raised due to several major events in Durban, according to Buckley (2017, 2): "the 2000 cholera outbreak; full-scale roll-out of free water, sanitation and hygiene education to unserved; and 60,000 full VIP toilets – how to empty?" The latter two problems were amplified by the 2000–2001 expansion of the Durban municipality to include a much larger peri-urban and rural section of what had formerly been an apartheid Bantustan (KwaZulu), where more than 1.5 million people were ill-served by the prior municipal governments. (This expansion was part of the restructuring of municipalities, which in 2000 were reduced in number from 842 to 284.)

Durban officials spent the subsequent decade grappling with how to roll out non-waterborne sanitation. By design, geographical discrimination associated with VIPs and UDs is striking, in a city whose spatial relationships to water were always color-coded. Black people living in low-income areas – "townships" – were generally given municipal bucket removal services instead of the flush toilets that whites enjoyed. Septic tanks were increasingly common in white areas, but in 1896, the first Sewerage Outfall Works was introduced using the London Pneumatic Sewerage System developed by Isaac Shone a dozen years earlier. But the sewage ran directly into the sea and it was only during the 1960s, according to an official citation of the city engineer's celebrated career, that treatment began:

> Don Macleod was responsible for the planning and construction of Durban's first four sewerage treatment works (there had been none before that and sewage was not treated properly until then) He also saw many suburbs of Durban get their first sewer connections and move off septic tanks... Part of his legendary work also focused on the implementation of these sewage systems in the black communities.
> (North Durban Presbyterian Church, 2015)

In fact, however, sewage systems were so slow to be installed in both the black townships and the peri-urban areas, that it was only in 1989 that even close-in Clairwood (an Indian residential area just a few minutes' drive from the central business district) was shifted from the bucket system to water-borne sanitation (Scott, 1994, 258). Two decades later, there were still 9270 homes suffering the bucket system (Amisi, Bond, Khumalo and Nojiyeza, 2008).

Neil Macleod worked at Durban Water and Sanitation under apartheid for half his career – from 1973 until 1994 (including when his father Don had become chief municipal engineer) – and led the unit from the mid-1990s. After 2000, when Durban expanded, Macleod was given inordinate praise for his 1997 Free Basic Water innovation: providing 6 kilolitres/household/month (a tariff reform subject to much dispute given that poor people's consumption actually declined by one third as the 6–10 kl/hh/month price soared). After numerous awards, noted Alex Loftus (2005, 188), "there has become something of a cult of the Durban example and the individual at its helm."

However, confidence in water delivery within the old Durban municipality boundaries gave Macleod (2008a, 2) confidence to geographically map out a "sanitation waterborne edge" (later "Urban Development Line"), beyond which a vast peripheral band of the municipality was considered too poor to justify laying sewage pipes and treatment facilities. At that point, in 2008, 60,000 VIPs were not being emptied because to do so was "uneconomic," reported Macleod (2008a, 7). Macleod (2008b) concluded,

> A piped sewerage system is not economically justifiable in rural areas, where the densities are too low, and in these areas onsite sanitation is the only viable option available. The rapid densification of the municipality has led to the run-off of untreated sewerage and polluted storm water into a number of rivers.

The first sentence in this quote is contradicted by the second, because no matter the subsidization involved, "rapid densification" should be the basis for running sewage pipes to even impoverished informal settlements, since there are potentials to realize "merit goods" and "public goods" such as health improvements (especially with cholera threats, diarrhoea and HIV/AIDS so prevalent), the potential for desegregation and cleaner rivers (Bond, 2002). Typically, the main official reason the municipality gives for not providing services to such residential areas is that doing so would entail legitimating their validity as permanent urban settlements, which officials are loathed to do given they were often gained in social struggles through illegal land invasions.

Nevertheless, the merits of UDs were overwhelming, according to Macleod (2008a, 8), in part because they used "minimum amounts of water, if at all." The UD is designed to *not* use any flushing mechanism. The capital cost of each UD toilet was (at the time) an average of $500, but municipal maintenance costs fell away because "emptying is the responsibility of the household, with entrepreneurs already offering their services at $4 per chamber emptied" (Macleod, 2008a, 8–9). Furthermore, Macleod (2008a, 8–11) claimed, "follow-up visits after construction have increased acceptance levels and emphasized the family's responsibilities for maintenance of the toilet. The period needed for follow ups extends to years" because of the need to "evaluate acceptance of the solution and to confirm that the hygiene messages have been internalized." This innovation most impressed a *Science* journalist (Koenig, 2008, 744), whose laudatory article termed UD the "best solution" for sanitation in Durban.[1]

At that point, in 2007, Buckley (2012, 26) estimated the backlog of homes without sanitation was 203,222 in informal settlements and 21,469 in "rural traditional" areas still within the city limits. He counted 87,207 UD toilets by then and an estimated 40,000 VIPs. As for flush toilets, 37,288 homes had septic tanks, and 498,341 had waterborne sanitation. Rounding up, in sum, of 912,000 Durban households, 535,000 had flush toilets; 225,000 had no sanitation; and 152,000 had non-flush toilets within 200 meters (Buckley 2012, 26).[2] The UD recipients all lived beyond "the sanitation waterborne edge," a vast peripheral band of the municipality in which it was deemed fiscally unrealistic to lay sewage lines; the UDs were not installed within the old limits, with the exception of Buckley's pilot at his residence.

By then, tens of thousands of VIPs had reached capacity, but the city was unable to fulfill its commitment to a "free basic sanitation service in the form of one pit emptying every five years," as Macleod (2008a, 7) conceded at the 2008 Africa Sanitation conference in Durban. Indeed, many pits were unlined with the "toilets subject to catastrophic collapse," many were "constructed in inaccessible locations," and there was high variability in content, size and cost of emptying. The cost of emptying each pit averaged $120 per pit, compared to "the cost of constructing new single Ventilated Improved Pit-latrine (VIP) type toilets: $140 to $420," making the process "uneconomic" (Macleod 2008a, 7).

The economics of sanitation remains subject to debate. In 2014, shortly after winning the world's premier prize for water services (the Stockholm Water Industry Award) and under attack from community activists, Macleod announced on an email list-serve that "the cost of providing water borne sanitation to the 370,000 families living in rural and shack areas would cost more than R60 billion" ($5.7 billion at the time) (Centre for Civil Society, 2014). That translated to R162,000 ($15,300) for each household, which appears far beyond a realistic range. Indeed, the construction cost of a new small "RDP house" with internal wet core was in the same range. In rural areas, he explained,

> it would cost 10 to 20 times more per household to construct sewers, compared to the denser urban areas. . . . In my view if families want more expensive options than that envisaged as basic sanitation, they should do what everyone else in the world (rich and poor alike) does and make a plan themselves instead of waiting for government.
> (Centre for Civil Society, 2014)

Moreover, Macleod continued, "The ongoing additional operational costs that would have to be funded from a revenue source would exceed R1.5 million a day" ($141,000/day), a more likely cost given the inability of those roughly 1.5 million people to afford full cost recovery (Centre for Civil Society, 2014). Regardless of how far off Macleod was in estimating capital costs, nevertheless it is clear that the main reason chosen not to supply waterborne sewage was the inability of households to cover its repayment.

The imperfect toilet, seen from below

In contrast, a critical school of thought starts from the households and communities themselves. Galvin (2017, 113) acknowledges, "There is little question that the introduction of UDs was handled very poorly. Officials explain that the education around use of UDs was organized through local councillors, who are their bosses and the democratically elected leaders. Councillors were tasked with outsourcing education." They didn't do so properly.

Several flaws associated with the "economically justifiable" denial of flush toilet systems were soon obvious. One is the blurred distinction between solids and liquids, given that diarrhoea often accompanies AIDS-related opportunistic diseases, and Durban has the world's highest level of urban HIV+ prevalence. Durban's extreme summer humidity for several months a year makes drying anything difficult. Most Durban townships and especially shack settlements suffer hilly terrain. Homes are often located on steep slopes, suffering poor drainage and facing often turbulent rain, leaving any structure – especially a small UD – vulnerable to landslides.

As a journalist (Veith, 2010) discovered after investigating these households' sanitation strategies, "As soon as they can afford it, people invest in a septic tank and abandon the dry toilets." In reply, predicted Macleod, the UD system would also benefit poor people once urine began to be collected and sold to the municipality (at $2.50/month) so as to dry it, since urine is "rich in nitrates, phosphorus and potassium, which can be turned into fertiliser" (Veith, 2010). He continued, "If we can turn the toilets into a source of revenues, then they will want to use the toilets." (That particular pilot project was never implemented at scale.)

These are all reasons that a larger social commitment to greater capital investments, as well as at least a low-flush sanitation system[3] – e.g. a well-maintained biogas anaerobic digester in each geographically-appropriate area (depending upon density, gradients and proximity to organic waste) – would have been the logical way for Durban to erase its "sanitation edge" to the point where there is no effective discrimination between households of different incomes, when it comes to removing faecal matter.[4] Since the late 1950s, anaerobic biogas digestion has been an option for low-flush sanitation in South Africa, including an appropriate-technology, pre-cast cement design that would apply to the kinds of urban shack-settlement and peri-urban residential patterns common to Durban (Mutungwazi, Mukuma and Makaka, 2018).[5] The only barrier to widespread implementation of this strategy is capital costs and political will.

For various reasons, complaints about UDs became increasingly common, including near the Inanda Dam. As community organizer Dudu Khumalo remarked about the Umzinyathi and KwaNgcolosi pilot communities, "These communities are repelled by human excrement as fertiliser, because of the many diseases surrounding them, compared to cow-dung. The burden of cleaning is left to women. Other creative opportunities for bio-gas are also foreclosed by UDs" (Amisi, Bond, Khumalo and Nojiyeza, 2008). Forms of everyday resistance soon emerged. One was to convert the UDs either to storehouses or as a top structure for self-dug septic tanks, supplied through new, informal piping that carries water to flush a dozen litres with a short distance to disposal.[6]

A techno-fix to the UDs' socio-technical failures and "perceived discrimination"?

Buckley's (2017) ongoing (self-)celebration of UDs contrasts with Macleod's eventual realism. To his credit, upon receiving the Stockholm Water Industry

Award in 2014, he conceded on the institution's website that the pipeless strategy entailed class discrimination (even if it was merely "perceived"):

> What we've realised is that into the future, we need to find new technologies that meet people's expectations. The reality is that everyone believes that the flushing toilet is the best solution to sanitation. . . we'll bring safe sanitation at an acceptable level to rich and poor alike and we'll do away with this perceived discrimination where the flushing toilet is seen to be for rich people and dry sanitation is seen to be a solution for poor people. Our challenge is to do away with that differentiation.
> (Centre for Civil Society, 2014)

However, *recognising* discrimination did not mean *changing* the system of "differentiation."[7] The UD was never installed in wealthier areas (aside from Buckley's house, to his credit), in spite of the claim to its universal merits. And in 2017, a rescue operation was mounted, based on another technical fix to the failed technical fix: an innovation (supported by Gates) to reform the UD so as to justify more regular collections of the excrement at the municipality's own cost. The new strategy was to supply UD faecal sludge to a South Durban (Isipingo) wastewater treatment plant by truck, so that two British entrepreneurs running a business on the municipal site, Agriprotein, could convert the sludge into a useful feedstock and then natural oils, using the larvae of several billion black soldier flies.

Durban's objective, according to Oxfam's Esther Shaylor (2018), was to set up a "faecal waste processing plant that produces beneficial products that ultimately reduce municipal running costs." This was an ideal market-based strategy in which a private firm would solve a problem of state-society frictions, Shaylor (2018) continued: "Due to health risks to householders, the Municipality decided to provide a waste removal service. The default option for disposal of the faecal waste was burial on site with tree-planting. However, there were concerns with environmental acceptability of burial onsite and space constraints in the long term." UKZN Pollution Research Group members Ellen Mutsakatira, Chris Buckley and Susan Mercer (2018, 2) offered different rationales for the same project: "In response to users' dissatisfaction at having to handle their own waste and evidence to show that sludge is potentially pathogenic after a year in a UDT vault, the municipality looked for alternatives to treatment of UD sludge."[8]

The chosen solution, as described by a BBC reporter (Gray, 2017) was to send the municipality's South Durban treatment works three tonnes of relatively dry UD faecal sludge daily (though it was unclear how many UDs this represented, given low levels of utilization). The sludge was "inoculated with young larvae before being harvested 13 days later" at which point the black soldier flies "will be able to generate up to 940 litres of oil a week from the waste it receives when the plant is fully up and running. The oil is sold as fuel but there could be other opportunities too – it is high in lauric acid, a compound commonly found in coconut oil and is often used in soaps and moisturisers" (Gray, 2017). An enormous biophysical effort would be required, so according to a CNN report, AgriProtein's Durban factory had, by mid-2018, a stock of 8.4 billion flies, consuming 276 tons of food waste daily, on which 340 million eggs were laid (Lo, 2018).

But two questions remained (not answered as this chapter went to press): would the flies' biophysical processes work as well with UD sludge as with other feedstock (such as food waste), and would the UD inputs be considered financially sustainable? On the first point, initial tests reported by Mutsakatira, Buckley and Mercer (2018, 4) based on a single Durban case experiment were fairly positive, showing the ability of the larvae "to reduce the UD sludge by 31 percent dry basis on a full-scale operation operating on an uncontrolled and low maintained system. . . comparable to literature values, which occurred at laboratory scale and in a controlled system." But they acknowledged that in other comparable studies, there was a much greater waste reduction ratio (e.g. from 50 to 79 percent) in part because of higher-nutrient inputs: "pit latrine sludge and fresh faeces on wet basis" and "other feed substrates such as chicken feed, market waste and municipal organic waste." They also admitted that given the relatively low rate they had achieved with Durban UD sludge, "a combination of different factors like the feeding rate, larval density and feed should be explored to see if the waste reduction and bioconversion can be increased."

As for the second point, Agriprotein co-founder Jason Drew implies that without recent subsidization by the Gates Foundation, the project would not have worked (Lo, 2018). The problem is similar to that associated with selling sanitation services everywhere. A private partner – who (unlike a state) does not benefit from the resulting merit and public goods, such as public health benefits from better sanitation, gender equality or desegregation – typically demands a high rate of return on their investment (30 percent of equity is normally what Foreign Direct Investment entails but this is before the devaluation of the local currency against hard currencies, which pushes the expected rates of return far higher). Drew conceded, "We spent nearly five years in abject failure. If I had known how hard it would be, and how much it would cost, I would probably not have started" (Lo, 2018).

In short, notwithstanding high-profile (BBC and CNN) promotion of the black soldier fly's miraculous faeces consumption capacity, neither the technical nor financial aspects of this innovation appear viable so far. The emptying and trucking of the UD sludge will probably continue to be erratic. And while techno-fixes to socio-technical problems are occasionally appropriate, and while any strategy to reduce water wastage in flushing without compromising dignity is welcome, as noted previously, it is a fair question as to whether relying on the black soldier fly to resolve the UD crisis is adequate. If it works, it will simply cement in a system that those at the lower tier of society, again and again, tell the municipality they do not appreciate.

Indeed, this innovation is reminiscent of another of Durban's high-profile, pilot-project piping investments that went sour: the 2000s installation of a waste-to-energy, methane-capture and generation plant in Durban's Clare

Estate community, funded by the World Bank via the Clean Development Mechanism. It failed for all manner of reasons, including technical shortcomings, financial over-estimations, environmental policy misinterpretations and social resistance (Bond, 2019).

Conclusion: eco-rhetorics and class realities

Durban's systemic failure to make – and maintain – adequate capital investments in water and especially sanitation was one reflection of the market-centric approach's profound weakness, which is relying on demand-led, weakly-subsidized systems where operating and maintenance expenses suffered from constant financial squeezes. Defenders of Durban's approach have never rebutted the central critique of the motivation lying behind the city's post-2000 sanitation policies: unnecessary fiscal neoliberalism. Instead, Macleod's post-retirement work has been dedicated to *obfuscating* that reality, e.g. with lectures based on progressive principles:

- a rights-based approach to water and sanitation services provision enshrined in the constitution with a subsidy policy that targets the provision of basic water and sanitation services to the poor
- a supportive political structure at the municipal level led by a mayor with a business-like approach to management in the public sector
- a clear separation of the oversight and management roles together with strong regulation at a national level
- "ring-fenced" financial accounts audited annually
- minimal pressure to engage in corruption or nepotism
- "learning by doing" in an organizational culture that encourages innovation, within clear risk boundaries (Macleod, 2016, 9).[9]

The next rhetorical step, for officials to justify low-quality sanitation services like the UD, is then quite logical: the increasing intensity of droughts. According to Gounden, "South Africa is a water-stressed country. With the increase in demand for drinking water, we cannot afford to flush this valuable resource down the sewer" (Veith, 2010). It is true that genuine water shortages are emerging, created not by neoliberal policy, inadequate fiscal support and racist planning – as in Durban – but instead by anthropomorphic climate change. Durban faces worsened humidity at certain times of the year, along with amplified dangers of extreme weather, including cyclones and storm surges, floods and droughts.[10] Durban's neoliberal strategy can then, indeed, sometimes be at least rhetorically justified on ecological-sanitation ("ecosan") grounds.

But the obvious rebuttal also arises: if water wastage is to be avoided, *then the logic of using eco-san principles in high-consumption areas is ecologically vital*, and hence the UD and other conservation measures should be tried first and most aggressively in the *wealthier* parts of Durban. Of course, if that were to happen, the city would face a fall in revenues, due to falling consumption at the high-income (water-hedonistic) end of the residential spectrum. Once again, the logic of the market, not good public policy, will shape water access.

It is here, finally, that we can invoke one of the most important thinkers at the interface of economic and environmental adaptation to consider how best to achieve steady-state, sustainable water and sanitation services, Herman Daly (1991, 19):

> It is absolutely a waste of time as well as morally backward to preach steady-state doctrines to underdeveloped countries before the overdeveloped countries have taken any measure to reduce either their own population growth or the growth of their per-capita resource consumption. Therefore, the steady-state paradigm must first be applied in the overdeveloped countries. . . . One of the major forces necessary to push the overdeveloped countries toward a. . . steady-state paradigm must be Third World outrage at their overconsumption.

Notes

1 Koenig (2008, 744) also remarked on how Macleod promised that by 2010, "everyone" within the expanded Durban city limits would have basic water and sanitation. Macleod (2008b) also promised, that year, "It can be reasonably expected that the housing backlog will be eradicated by 2015" – although in April 2019 it was admitted by municipal officials that this backlog had lengthened and there would be a housing shortage "for forty years" (Pillay, 2019).
2 VIPs were increasingly filled up, but not being emptied, and as former Johannesburg water regulator Kathy Eales remarked: "Many VIPs are now full and unusable. In many areas, VIPs are now called 'full-ups'. Some pits were too small, or were fully sealed." Public policy, she conceded, "does not clarify roles and responsibilities around what to do when pits are full" (Amisi, Bond, Khumalo and Nojiyeza, 2008).
3 The typical 9 litre South African flush is required to carry small amounts of solid excrement many kilometres, via pumping stations and sewage treatment plants, back into water systems for reuse. An alternative – for more geographically-proximate disposal – is the ever-improving "micro-flush" strategy (e.g. 2 litres that can flush out solids, available in various new designs, including from the Gates Foundation). Again, this requires a higher initial capital investment in disposal systems such as septic tanks, soakaways, and biogas digesters.
4 UN Habitat's recommended low-cost sanitation system has various advantages over the UD system, and has witnessed more than a million installations in India: "The twin-pit system uses 1.5–2 litres of water per use in a flush toilet that is connected to two pits that allows recharging of the soil and composting, and a close-loop public toilet system attached to a bio-gas digester. In fact, this is the only sanitation technology that meets the seven conditions for a sanitary latrine laid down by the World Health Organisation. These stipulate that a sanitary latrine should not contaminate surface soil, ground water or surface water. Excreta should not be accessible to flies or animals. There should be no handling of fresh excreta, or when this is unavoidable, kept to a bare minimum. There should be no odor or unsightliness and the methods used should be simple and inexpensive in construction and operation" (Reddy, 2007).

5 In 2009, Durban witnessed a high-profile biogas digester pilot project disaster, due to a municipal/NGO project's failure to consult with Cato Manor residents about what they denigrated as a "shit fish tank" (Desai, 2011).
6 That, in turn, created new dilemmas for the strategy increasingly known as "commoning," in which communities find ways to illegally reconnect water and electricity so as to avoid punitive pricing and excessively expensive cost recovery. In the area near Inanda Dam, given the very weak pumping systems installed by the municipality, the illegal connection of water to new residences meant that at the end of the pipes, there was insufficient pressure to deliver water (Bond and Galvin, 2019).
7 This is also recognized at the national scale, e.g. by long-serving Water Research Commission director Jay Baghwan (2019): "The binary model, of the gold standard in the form of full flush toilet vs hole standard in the form of pit latrines, for rich and poor areas respectively, has not closed the gap but created a myriad of new operational challenges. Compounding this is the fact that South Africa is a water-scarce country and the universal access to waterborne sanitation may not be realized due to the prohibitive costs and the availability of water. The deeper problem is that there is no sanitation market, especially for the poor – it happens to be a monopolized public good with minimal innovation uptake."
8 Typically, faecal sludge stays pathogenic 12–18 months after burial, with 10–25 percent of bacterial eggs retaining pathogens.
9 Macleod (2016, 11) conceded that municipal managers had learned that "expectations from a toilet" included "a place of refuge and contemplation; a place to be private and read, sing, cry, etc; a clean place that does not smell; safe to use; and private, clean, odor free, well lit and safe to access and use." What Macleod failed to mention was the finding from 17,000 surveys that nearly everyone surveyed *expected to be able to flush*, and this was the main dignity-based rationale given for the rejection of his UD strategy.
10 Although an October 2017 superstorm showed that flood management requires much more work, Durban's bulk water supply is presently satisfactory thanks to (belated) capital investment which doubled inflows on the city's south side via new pipes from Spring Grove and Midmar dams, and, on the drought-prone north side, doubled the capacity of Hazelmere dam by raising its wall height (from 2014–2018 Hazelmere suffered extremely low water levels). But granted, periodic droughts should compel a rethink of the present 9-litre-per-flush system, amongst other ways to save water. And as noted previously, an ideal replacement, over time, with no discriminatory implications, would be the conversion of these high-volume flush toilets and connector infrastructures, into lower-volume (e.g. 2 litre), locally-oriented biogas digestive chambers with catchments large enough to reach economies of scale.

References

Amisi, B., Bond, P., Khumalo, D. and Nojiyeza, S. (2013) 'The neoliberal loo.' Available at: http://ccs.ukzn.ac.za/default.asp?3,28,11,4173
Bond, P. (2002) *Unsustainable South Africa*. London: Merlin Press.
Bond, P. (2013) 'Water rights, commons and advocacy narratives,' *South African Journal of Human Rights*, 29 (1), pp. 126–144.
Bond, P. (2019) 'The BRICS New Development Bank's false dawn,' in Shaw, A. and Bhattacharya, R. (eds) *Urban housing, livelihoods and environmental challenges in emerging market economies*. New Delhi: Orient Blackswan.
Bond, P. and Galvin, M. (2019) 'Water, food and climate commoning in South African cities,' in Vivero-Pol, J.L. and Ferrando, T. (eds) *Routledge handbook of food as a commons*. London: Routledge. Available at: www.routledge.com/Routledge-Ha

ndbook-of-Food-as-a-Commons/Vivero-Pol-Ferrando-Schutter-Mattei/p/book/9781138062627

Buckley, C. (2017) 'Urine diversion in Durban.' Håkan Jönsson Farewell Symposium, October 25, Stockholm. Available at: http://blogg.slu.se/kretsloppsteknik/files/2017/10/UKZN-for-Hakan-V10.pdf

Centre for Civil Society (2014) 'Durban Water angers activists, impresses Stockholm judges.' Available at: http://ccs.ukzn.ac.za/default.asp?2,68,3,3308

Daly, H. (1991) *Steady state economics*. Washington, DC: Island Press.

Desai, A. (2011) 'Where eco wars are truly felt,' *The Mercury*, November 29.

Galvin, M. (2017) 'Leaving boxes behind: Civil society and water and sanitation struggles in Durban, South Africa,' *Transformation*, 92, pp. 112–134. Available at: https://muse.jhu.edu/article/648943/summary

Gates, B. (2010) 'Simple advances, amazing benefits in Africa,' *The Gates Notes*, March 18. Available at: www.thegatesnotes.com/Topics/Health/Simple-Advances-Amazing-Benefits-in-Africa

Gounden, T., Pfaff, B., Macleod, N. and Buckley, C. (2013) 'Tackling Durban's sanitation crisis head on,' *Infrastructure News*, January 9. Available at: www.infrastructurene.ws/2013/01/09/tackling-durbans-sanitation-crisis-head-on

Gray, R. (2017) '"Fatbergs", faeces and other waste we flush could be a fuel,' *BBC*, October 5. Available at: www.bbc.com/future/story/20171005-human-waste-could-be-the-fuel-of-the-future

Kaye, L. (2012) 'What is the future of toilet technology?' *The Guardian*, October 9. Available at: www.theguardian.com/sustainable-business/future-toilet-technology-sanitation-water

Koenig, R. (2008) 'Durban's poor get water services long denied,' *Science*, 319, pp. 744–748.

Lo, A. (2018) 'Two brothers want to revolutionize the food industry with maggots,' *CNN*, September 27. Available at: https://edition.cnn.com/2018/09/27/business/agriprotein-fly-farming/index.html

Loftus, A. (2005) '"Free water" as commodity,' in MacDonald, D. and Ruiters, G. (eds) *The age of the commodity*. London: Earthscan.

Macleod, N. (2008a) "Reaching low income communities: Experiences from eThekwini, Durban." AfricaSan conference, February, Durban, South Africa. Available at: www.slideserve.com/loren/africasan-reaching-low-income-communities-experiences-from-ethekwini-durban-powerpoint-ppt-presentation

Macleod, N. (2008b) 'We are committed to cleaner rivers ,' *The Mercury*, February 19.

Macleod, N. (2016) 'Turning around under-performing water and sanitation utilities and targeting the poor.' Available at: http://sanitationandwaterforall.org/wp-content/uploads/download-manager-files/NeilMacleodTurning%20around%20under-performing%20water%20and%20sanitation%20utilities%20and.pptx

Merriam-Webster Sewage.' Available at: www.merriam-webster.com/dictionary/sewage larvae in faecal sludge management." 41st WEDC International Conference, July 9–13, Egerton University, Nakuru, Kenya. Available at: https://dspace.lboro.ac.uk/dspace-jspui/bitstream/2134/35872/1/Mutsakatira-2994.pdf

North Durban Presbyterian Church (2015) 'Don honoured,' August 30. Available at: www.ndpchurch.co.za/?p=966

Pillay, K. (2019) 'City to buy, expropriate land for housing,' *The Mercury*, April 17.

Reddy, A.P. (2007) 'World Toilet Summit: Sanitation beyond septic tanks and sewers,' *InterPress Service*, November 6. Available at: www.ipsnews.net/2007/11/environment-sanitation-beyond-septic-tanks-and-sewers

Republic of South Africa (2001) *White paper on basic household sanitation.* Pretoria: Department of Water Affairs. Available at: www.dwaf.gov.za/Documents/Policies/Sanitation Review Policy_.doc

Republic of South Africa (2002) *Regulations under Section 10 of the Water Services Act (Act No. 108 of 1997): Norms and Standards for Water Services Tariffs (Explanatory Notes and Guidelines).* Pretoria: Department of Water Affairs. Available at: www.dwa.gov.za/Dir_WS/wspd/UserControls/DownloadImportFiles.aspx?FileID=177

Scott, D. (1994) *Communal space construction.* Doctoral thesis, University of Natal Department of Geography and Environmental Engineering, Durban. Available at: https://researchspace.ukzn.ac.za/bitstream/handle/10413/7364/Scott_Dianne_1994.pdf?sequence=1&isAllowed=y

Shaylor, E. (2018) 'SuSanA webinar: The use of Black Soldier fly larva for faecal sludge treatment,' January 18. Available at: https://forum.susana.org/147-production-of-insect-biomass-from-excreta-or-organic-waste/22101-susana-webinar-the-use-of-black-soldier-fly-larva-for-faecal-sludge-treatment-30th-january-2018

South African Human Rights Commission (2018) *The right to water and sanitation.* Available at: www.sahrc.org.za/home/21/files/SAHRC%20Water%20and%20Sanitation%20revised%20pamphlet%2020%20March%202018.pdf

Veith, M. (2010) 'Urine for sale? Durban's in the market,' *Mail & Guardian*, November 7. Available at: https://mg.co.za/article/2010-11-07-urine-for-sale-durbans-in-the-market

Index

Note: page references in bold indicate tables; 'n' indicates chapter notes.

Abers, R.A. 31, 35
Aboriginal communities *see* Indigenous communities
accountability xv, 43, 69, 76, 105, 178
accumulation by dispossession: bottled water as 115–17, 119, 124n1; water privatization as 131–32, 134–35, 185
Africa 114, 149; *see also* Middle East and North Africa (MENA); South Africa; Tanzania
agrarian *sindicatos* (peasant unions) 102, 110n4
AgriProtein (company) 196–97
Alatout, S. 57
Albert, G. 47
Allen, A. 101, 102
American Civil Liberties Union (ACLU) 45
Anand, N. 17, 32
Anand, P. 71
Angel, J. 5, 85, 182
Anheuser-Busch Brewery 121
anthropocentrism 47–50
Archambault, D., II 45–46
Archontopoulos, G. 164, 166
Arendt, H. 15–16, 23
Argentina 31, 145–46
Aristotle 56
artificial water scarcity 191
Athens, Greece, water privatization in 161, 163–67; 'Save Greek Water' 166
austerity policies 133, 138, 144, 161–64, 169, 171, 176–77

Baer, M. 31, 34
Baghwan, J. 200n7
Bakker, K. 10, 18, 116
Bangladesh 92

Barcelona, Spain 149–50
Barlow, M. 1
Barnes, J. 57
Bechtel (company) 109n3, 152
belief 63
Bello, W. 9
Bergson, H. 18
Berlusconi, S. 136
Bieler, A. 169
biocontrol 122, 124n5
Blaser, M. 56, 59, 60–61, 65n6
Boelens, R. 91, 118
Böhm, S. 118
Bolivia 4, 30, 78, 92; informal settlements in 99–109; Law of Popular Participation (LPP; 1994) 105, 107; water remunicipalization in 152–153; *see also* Cochabamba
Bonafont (brand) 117
Bond, P. 184
Bos, J.J. 87
bottled water 10–11, 35, 113–24; as accumulation by dispossession 115–17, 119, 124n1; ethics of 117–20; implications for the right to water 123–24; supply to disaster areas 120–23, **121**
boxed water 118
Brazil 31, 90
Bread for the World (NGO) 77
Brei, V. 118
Brooks, D. 78
Buckley, C. 191–93, 195, 197
Budds, J. 42
Buenos Aires, Argentina 145–46
Bustamante, R. 4, 100

Canada 18, 55, 57–59, 63, 64, 145
canned water 115, 118, 121–22

Index

capabilities approach 71–72, 85, 86; *see also* water-security capabilities approach
Cape Town, South Africa, sanitation in 190
capitalism 48, 51, 129–35, 139–40, 170, 185; racial 9
Catholic Church 43, 49–51, 108, 137
Chakrabortty, A. 132
Charpleix, L. 47
Chile 30
civic participation 184–85
civil society organizations xiv, 5, 34, 37–38, 108, 165, 182
Clark, C. 4, 5, 9
class 7–8, 103, 130–31, 134, 169, 198–99
Cleaver, H. 116, 129, 133, 134
clientelism 38
climate change 3, 18, 21, 22, 135, 198
Cochabamba, Bolivia 135; Association of Communitarian Water Systems and EPSAS of the Zona Sur (ASCIASUDD-EPSAS) 107–8; formal and informal water providers in 102–3; informal urbanization in 103–5; SEMAPA (public water utility) 102, 104, 108; Water War 100, 101, 105–6, 108, 109n3, 152, 153
co-gestion model 108
Colombia 31
coloniality 7–9, 18, 43–46, 49, 51, 64
commodification of water 10–11, 68, 116–19, 122; *see also* marketization of water governance
commoning 200n6
commons, the 11, 60, 115, 119, 139, 167–68
communities 23–24, 90, 92–93, 99–108; *see also* cooperatives; coproduction partnerships; Indigenous communities; informal water providers
Comprehensive Economic and Trade Agreement (CETA) 138
constitutive financialization 178
cooperatives 102, 106, 139–40, 165, 171
coproduction partnerships 101–2, 108
corporate social responsibility 118, 121–23
Crespo, C. 4, 100
cultural-religious traditions 8, 42–51, 90; Catholic Church 43, 49–51; ontologies of water and 63; and the right to water declaration 48–49; Standing Rock, North Dakota 43–46; Whanganui river, New Zealand 43, 46–48

Dakota Access Pipeline (DAPL) 44–46
Daly, H. 199
dams 108, 109n3, 110n7, 191, 195, 200n6, 200n10
Danone (company) 117
Dávila, G. 91
de Albuquerque, C. 3–4, 34
debt 132, 138, 178, 180
decentralization 105, 109n3
Declaration of the Rights of Mother Earth 48–49
decolonization 64
della Porta, D. 133
democracy 139–40, 178, 185
Democratic State Government Representatives, Michigan 184
Descartes, R. 55
Detroit, Michigan, water crisis in 175–85; background 175–77; Emergency Management 176–79; equality 179–80; financialization, water and the right to the city 183–85; role of governance 177–79; roles of the state 182–83; socioeconomic rights and 180–82
Detroit Water and Sewerage Department (DWSD) 177
dignity, human 15, 68, 78, 85, 86, 93, 197, 200n9
disaster zones, bottled water supply in 120–23, **121**
disconnections, water supply 4, 90, 175–77, 179–80, 182–84
discrimination 192–93, 195–98, 200n7; *see also* non-discrimination
domestic water 15, 30, 38, 48, 68–69, 73, 76–78; *see also* water for life
donkeys for water haulage 28
Draghi, M. 138, 163
Drew, J. 197
drought 190, 198, 200n10
Durban, South Africa, sanitation in 189–99, 200nn5–7; Free Basic Water 72, 191, 192; Urine Diversion (UD) toilets ("neoliberal loo") 190–98, 199n4, 200n9
Durban Water and Sanitation 192
duty 89–90
Dwinell, A. 182

Eales, K. 199n2
Eckstein, G. 19
ecological-sanitation ("ecosan") principles 198–99
ecological security 6, 7; *see also* environmental protection

economic crisis 135–38; *see also* financial crisis (2007–2008)
economics 20–23, 115–17
ecosystems 76, 89
Ecuador 30, 78
education 92
Emergency Management 176–79
Energy Transfer Partners 45
entrepreneurialism 178, 196
environmental protection 134–35; *see also* ecological security
environmental racism 179
equality 15, 19, 23, 100, 176, 179–80, 184–85; *see also* inequality
equity 32, 91–94, 139, 145, 149, 181; *see also* inequity
Erdrich, L. 46
Escobar, A. 60
ethics 42, 43, 45, 48, 50, 51, 87, 117–20
Ethos (brand) 118
European Central Bank 132–33, 138, 161, 163
European Citizens' Initiative, 'Right2Water' 5, 129, 136–38, 150–51, 161–63, 166, 169–72
European Commission 132–33, 138
European Federation of Public Services Unions (EPSU) 167
European public water 129–40; agency of resistance and 133–35; and economic crisis 135–38; privatization and 131–33; structure and agency in 130–31; water remunicipalization in 149; *see also* France; Germany; Greece; Hungary; Ireland; Italy; Portugal; Spain; Switzerland; United Kingdom
European Union (EU) 5, 108, 136, 161
expert pressure 32–35, **33**

financial crisis (2007–2008) 20, 129, 132–33, 163
financialization of water 183–85
First Nations peoples 55, 57–59, 63, 64
Flint, Michigan, water crisis in 8, 9, 23–24, 92, 175–85; background 175–77; Emergency Management 176–79; equality 179–80; financialization and the right to the city 183–85; role of governance 177–79; role of the state 182–83; socioeconomic rights and 180–82
flow control 116
flush toilets 189–93, 195–98, 199n3, 200nn9–10
food security *see* right to food
food sovereignty 77

Foucault, M. 122
France 132, 144, 150
Francis, Pope 34, 35, 50, 84
Franco, J.C. 68
Fraser, N. 134–35
Frenk, G.A. 154
fulfillment 70, 71, 73, 181

Galvin, M. 191, 194
Gates, B. 191
gender 7, 134–35; *see also* women
Gerber (brand) 117
Gerlak, A. 31, 34
Germany 136–37
Gimelli, F.M. 87
globalization 133
global North 2, 3, 9–11, 175
global South 2, 7–11, 20, 38, 69, 99, 101, 175, 179, 185
Global Water Partnership 20
Goldin, J. 87, 91
Gounden, T. 191, 198
Gramsci, A. 134
Greece, citizen mobilization against privatization in 129, 137–39, 161–72
"green economy" concept 20–21
Grey, D. 6, 85
groundwater 90, 101, 102, 106
Guardian, The (newspaper) 47, 132, 191
Guatemala 30

Hall, R.P. 69, 75
Hamilton, Canada 145
Harris, L. 8, 57, 61, 64, 116, 180, 182
Hart, G. 9
Harvey, D. 119, 131, 178, 184
Hazelmere dam 200n10
hedging 35–36
Heller, L. 28, 34
Hellum, A. 68, 78
Heywood, P. 60, 62
Hopi tribe 93
Horváth, T.M. 147
human rights 15; non-contingent nature of 16–19
Human Rights Watch 30
Humphrey, J. 18
Hungary 147
Hurricane Harvey and bottled water consumption 120–23, **121**
hydrosocial cycle 54

Inanda Dam 191, 195, 200n6
India 17, 32, 199n4

indigeneity 7–8, 21, 43, 45, 48, 103; *see also* cultural-religious traditions
Indigenous communities 18, 43–51, 55, 60, 70, 86, 101
Indonesia 119, 151
inequality 6–8, 19, 22–24, 145, 175; *see also* equality
inequity 91–92, 11; *see also* equity
informal water providers 28, 34, 35, 37, 99–105, 107, 124
infrastructural coexistence 37–39
injustice 7, 118–19; *see also* justice
Inkster, J. 58
Inouye, D. 44–45
insecurity, water 6, 7, 28, 30, 34, 124
institutions 3–6, 29–32, 36, 101, 103, 106
integrated water resources management (IWRM) 19, 20
international law 75, 143–44
International Monetary Fund (IMF) 109n3, 132–33, 138, 161–163
investment funds 131–32
Iraq 44
Ireland 163
irrigation 69, 103, 110n4, 110n7
Isla Urbana, Mexico City (civil society organization) 37–39
Italian Water Forum 136
Italy, water privatization in 129, 135, 137–39, 163, 167; Comitato Italiano Contratto Mondiale sull'Acqua (CICMA) 136

Jakarta, Indonesia 151
Jepson, W. 64, 72, 87, 89, 90, 180, 182
Johnson, L.B. 176
Johnson, T. 44
justice, water 2, 86, 118, 175, 182, 183; *see also* injustice

Kameri-Mbote, P. 78
Karyotis, T. 168
Kazakhstan 147
Keck, M. 31, 35
Kipfer, S. 9
knowledge systems 43–44
Koenig, R. 199n1
Konstantopoulou, Z. 170
Kysar, D. 21

land rights *see* property rights
Landry, C. 45
language of human rights 182–83

Latin America 30–31, 99, 101, 105, 149; *see also* Argentina; Bolivia; Brazil; Chile; Colombia; Ecuador; Guatemala; Mexico; Peru; Uruguay; Venezuela
Latour, B. 60
Lavoisier, A.L. 56, 58, 59
Law, J. 55, 56, 61
Ledo, C. 102
Lefebvre, H. 184
liberalism 148, 179
Linton, A. 42
Linton, J. 88, 139
literacy 92
Loftus, A. 5, 85, 119, 144, 145, 154, 182–184, 192
loteadoras 104, 110n5
Lyda et al. v City of Detroit (2014) 182–83

Macleod, N. 191–96, 198, 199n1, 200n9
Macron, E. 171
Madrid, Spain 149–50
Mandamin, J. 58–59, 63, 65n5
Māori people 43, 46–48, 59
Maputo Protocol 78
marketization of water governance 30, 31, 90, 116, 118–19, 146–47
Marston, A.J. 107
Marx, K. 132
Marxism 134
Mazibuko v Johannesburg Water (2009) 190
McDonagh, S. 49
Meehan, K. 90
megaprojects xiv
Mehta, L. 68, 71, 87
Menchú, R. 91
Mercer, S. 196, 197
meters, prepaid water 179–80, 190
Mexican Congress 31–32
Mexico 28–39, 90; bottled water in 113, 119, 120; Centro Mexicano de Derecho Ambiental (CEMDA) 30; Citizen's National Water Law 36, 153–54; Coordinadora Nacional Agua Para Tod@s, Mexico (water activists) 36; expert pressure 32–35, 33; hedging 35–36; institutionalizing the human right to water in 30–31; National Water Law 32, 34, 35; practical authority for the right to water 29–39; practical experimentation 37–39; water remunicipalization in 153–54; Water Sustainability Law 39

Mexico City 28, 35–39; 19S earthquake and bottled water consumption 120, **121**, 123; Isla Urbana (civil society organization) 37–39; Sacmex (Sistema de Aguas de la Ciudad de México) 36
Middle East and North Africa (MENA) 78
Midmar dam 200n10
militarism 44–45
MillerCoors (company) 121
Misicuni dam 108, 109n3, 110n7
Mitlin, D. 101, 102
Mol, A. 54, 56, 65n6
Morales, E. 106, 107, 108, 152
moral luck 15–17, 19, 21–24
"Mother Earth Water Walk", North American Great Lakes 58–59, 65n5
Moyn, S. 19
Mumbai, India 17, 32
Murphy, C. 44
Mutsakatira, E. 197

Nadasdy, P. 21
nationalism 134–35
national security 6
negative rights 17
neoliberalism xiv, 30–31, 43, 105, 124, 129, 145, 149, 154, 162, 198; *see also* Urine Diversion (UD) toilets ("neoliberal loo")
Nestlé (multinational) 117
New Zealand 43, 46–48, 59
#NODAPL 45
non-discrimination 180–82
non-government organizations (NGOs) 10, 106–108, 136, 144, 155, 191
non-retrogression 181–82
non-state actors 10, 29, 31, 37–39
Nussbaum, M. 86–87

Obama, B. 46
obligation 89–90
Occupy movement 184
Olivera, M. 182
one-world world 55–56, 64
ontological conjunctures 61
ontological exteriority 130
ontologies of water 8, 54–64, 65n3; the ontological turn and multiple ontologies of water 55–59, 64; political 59–61, 63, 89; the right to (enact) water 61–63; and a water-security capabilities approach 89

Pacheco-Vega, R. 116
Paris, France, *Eau de Paris* 144, 150

Paris Commune (1871) 184
Parks, L. 133
Peck, J. 178
Perreault, T. 107
Peru 30
Pfaff, B. 191
philosophy of internal relations 130
planetary boundaries 20–21
Polanyi, K. 133
political will 69, 73
Portugal, water privatization in 161–163, 168
positive rights 17, 93
poverty 9, 72, 87, 119, 175, 180, 181, 183
power 7, 11, 47, 54–55, 69, 116, 122, 124n5, 134, 179
practical authority for the right to water 29, 39; establishing 31–32; expert pressure 32–35; hedging 35–36; practical experimentation 37–39
prepaid water meters 179–80, 190
privatization of water 4, 11, 30, 34–36, 46–47, 68, 72, 116–18, 124, 129, 139, 143–44, 163; resistance to 129, 133–38, 161–62, 165–68; structure of 131–33; *see also* remunicipalization
production, social relations of 130–31, 134–35
production, water for 69, 72, 75–78
profit motive 4, 36, 45, 101, 119, 124, 131–32, 139, 162
property rights 18–19, 103
protection 70, 73, 181
publicness theory 117
public-private partnerships 178
public-public partnerships (PUPs) 155
public service approach 117, 124

quality control 116

Rabasa, A. 32
race 7–9, 24, 134–35, 175–80, 183, 185, 190, 192
racial capitalism 9
rainwater harvesting 37–39, 90
Ranganathan, M. 9, 179, 180, 185
reality 55–56, 64, 65n3
religious traditions *see* cultural-religious traditions
remunicipalization of water 10, 143–55; anti-capitalism 152–53; autocratic state capitalism 146–47; autonomism 153–54; future prospects for 154–55; and the human right to water 145–54; market

managerialism 147–49; social democracy 149–51
resilience 19–24
respect 58–59, 62–63, 65n5, 70, 73, 181
rhetoric 198–99
right to bring waters into being *see* ontology of water
right to food and the right to water 5, 7, 68–79; basic tenets of 70–73; bridging the rights 77–79; convergences and possible tensions 75–77; latent potentials of 73–75
"right to have rights" concept 16, 23–24
right to sanitation xiii–xv *see* sanitation
right to water xiii–xv, 1–3, 11–12; basic tenets of 71–73, 80n2; between global South and global North 9–11; institutions and the state 3–6; latent potentials within 73–75; race/class/indigeneity/ coloniality and 7–8; water security discourses 6–7
rivers in legal frameworks 47–48, 59
rights to the city 183–85
Robinson, J. 46
Rogers, B.C. 87
Ruru, J. 47

Sadoff, C.W. 6, 85
San Francisco Peaks (*Nuvatukya'ovi*) conflict, USA 93
sanitation in Durban, South Africa 189–99; flush toilets 189–93, 195–98, 199n3, 200nn9–10; Urine Diversion (UD) toilets ("neoliberal loo") 190–98, 199n4, 200n9
scarcity, water, artificial 191
Sen, A. 71–72, 86
septic tanks 191–193, 195
settler colonialism *see* coloniality
Shaylor, E. 196
shortages, water 28, 198, 200n10
Shue, H. 17
Sioux (Great Sioux Nation) 44–46, 48
social movement unionism 167
socioeconomic rights 181
Sotiris, P. 133
South Africa 4, 92; water disconnections in 179–80, 182, 183; Water Services Act (1997) 189; *see also* Cape Town, sanitation in; Durban, sanitation in
Spain 149–50
Spring Grove dam 200n10
Standing Rock, North Dakota 43–46
Starbucks (company) 118

state, the 3–6, 8, 17–18, 70–79, 85, 89–91, 116, 182–83; informal settlements and 99–102, 106, 108
Stengers, I. 60
Strang, V. 47
Suez (company) 132, 143, 145–46, 150, 171
sufficiency 15, 16, 19, 22–24, 73
Sultana, F. 91, 119, 122, 124n5, 144, 145, 154, 183, 184
"superhero" solution 121, **121**, 124
sustainability 20–21, 23, 70–71, 76
Switzerland 78

Tanzania 148–49
tap water 113, 117, 120
Thames Water (company) 132
Thessaloniki, Greece, citizen mobilization against privatization in 161–72; 'Initiative 136' (K136) 165–66, 167, 168, 171, 172; 'SOSteToNero' (SOS for water) 166, 171, 172
Thompson, H.A. 176
toilets *see* sanitation
traditions *see* cultural-religious traditions
Trichet, J.-C. 138, 163
Troika 132–33, 138, 161–164, 169, 171
trucked water 28, 102
Tuck, E. 64

UN Committee on Economic, Social and Cultural Rights (UNCESCR) 70, 75, 181
UN Declaration on the Rights of Indigenous Peoples (UNDRIP) 46
UN Declaration on the Rights of Peasants and Other People Working in Rural Areas 77
UN Habitat 199n4
UN High-Level Panel of Experts on Food Security and Nutrition (HLPE) 5, 69, 78
UN High-Level Panel on Water (HLPW) 5, 15–24; "valuing water" initiative 5, 15, 16, 21–23, 24
UN Human Right to Water and Sanitation (HRtWS) xiii–xiv, 15, 18; *see also* right to water; sanitation
unilateral utility model 99
United Kingdom 132
United States of America 18, 144, 149, 175; *see also* Detroit, Michigan; Flint, Michigan; Hurricane Harvey; Standing Rock, North Dakota
UN Millennium Development Goals (MDGs) 3

UN Office of the High Commissioner for Human Rights 143, 181
UN Special Rapporteur for the Human Right to Water and Sanitation xiii–xv, 3, 5–6, 28–29, 34–35, 69, 76–77, 78, 99, 144, 182–83
UN Sustainable Development Goals (SDGs) 3, 10, 15, 16, 19, 21, 22, 23, 109
UN Universal Declaration of Human Rights (UDHR) 15, 16, 18, 24, 62, 65n7, 68, 70
UN Water 144
urban areas 100–105, 107, 177–79
Urine Diversion (UD) toilets ("neoliberal loo") 190–98, 199n4, 200n9
Uruguay 30, 151
US Army Corps of Engineers 44–45

Vance, E. 69
Van Houweling, E. 69, 75
Van Koppen, B. 68, 75, 77, 78
Veilleux, J. 45
Veldwisch, G.J. 68
Venezuela 30
Venkatachalam, L. 124
Ventilated Improved Pit-latrines (VIPs) 191–94, 199n2
Veolia (company) 132, 143, 150
Vía Campesina, La (organization) 77
violence, systemic 34

Walnycki, A.M. 4, 100
Wateau, F. 42

WaterAid 144
water-as-lifeblood 57–58, 61, 63
water for life 69; *see also* domestic water
water for production 69, 72, 75–78
water-security capabilities approach 6–7, 84–94, 182; emergence of 86–88; equity and 91–93; future of 93–94; revisioning the right to water through 88–91
water security discourses 6–7, 85, 88
well-being 31, 68, 72, 79, 86–89, 93–94
wells, collective 90
Whanganui River, Aotearoa/New Zealand 43, 46–48, 59
White Bull, B. 45
Whiteside, H. 178
Wilson, N.J. 8, 57, 58, 61
Windfuhr, M. 76
women 69, 76, 77, 92 see gender
World Bank 22, 109n3, 148, 149, 155, 162, 198
World Economic Forum 20
World Health Organization 72, 199n4
World People's Conference on Climate Change, Bolivia 48–49
Wutich, A. 64, 91, 92, 124, 180, 182

Yang, K.W. 64
Yates, J.S. 8, 57, 61, 180

Ziganshina, D. 77
Zimbabwe 116
Zwarteveen, M.Z. 118